Geometry For Dummies®

BESTSELLING BOOK SERIES

W9-BRU-404

Geometry Symbols

$\angle$, $\measuredangle$	Angle, angles	AB	Line AB
m$\angle$AB	Measure of angle AB	$\overline{AB}$	Line segment AB
$\overset{\frown}{AB}$	Arc AB	$\overrightarrow{AB}$	Ray AB
$m\overset{\frown}{AB}$	Measure of arc AB	$\parallel$	Parallel
$\odot$, $\odot$s	Circle, circles	$\nparallel$	Not parallel
$\cong$	Congruent	$\square$	Parallelogram
$\ncong$	Not congruent	$\perp$	Perpendicular
°	Degree	$\sim$	Similar
=	Equal	$\triangle$	Triangle
$\neq$	Not equal	(x, y)	Ordered pair in plane
>	Greater than	$\frac{a}{b} = \frac{c}{d}$	Proportion
<	Less than	$a{:}b$ or $\frac{a}{b}$	Ratio
AB	Length of line segment AB		

Famous Abbreviations

AA	Angle-angle, for proving triangles similar
AAS	Angle-angle-side, for proving triangles congruent
ASA	Angle-side-angle, for proving triangles congruent
cos	Cosine
cot	Cotangent
CPCTC	Congruent parts of congruent triangles are congruent
csc	Cosecant
CSSTP	Corresponding sides of similar triangles are proportional
SAS	Side-angle-side, for proving triangles congruent
sin	Sine
tan	Tangent

For Dummies: Bestselling Book Series for Beginners

Geometry For Dummies®

Cheat Sheet

Commonly Used Variables

a	Apothem
a, b, c	Lengths of the sides of a triangle
A	Area of a polygon
B	Area of the base of a solid
C	Circumference of a circle
h	Height of an altitude
α	Alpha (name of an angle)
β	Beta (name of an angle)
χ, x	Unknown value
θ	Theta (name of an angle)
π	Pi
ℓ	Slant length of a side of a solid
l or l	Length of a rectangle

L	Lateral area of a solid
m	Slope of a line
M	Midpoint of a line segment
n	Number of sides of a polygon
P	Perimeter of a polygon
P or P'	Plane
r	Radius
s	Length of the side of an equilateral polygon
S	Surface area of a solid
T	Total area
V	Volume
w or w	Width of a rectangle

Handy-Dandy Formulae

Area (A) of a triangle	$A = \frac{1}{2}bh$ where b measures the base and h the altitude
Perimeter (P) of a triangle	$P = a + b + c$ where a, b, and c are the lengths of the sides
Area (A) of a rectangle	$A = lw$ where l measures the length and w the width
Perimeter (P) of a rectangle	$P = 2b + 2h$ where b measures the width and h the height
Area (A) of a circle	$A = \pi r^2$ where r measures the radius
Circumference (C) of a circle	$C = 2\pi r$ or $C = \pi d$ where r measures the radius and d the diameter

For Dummies: Bestselling Book Series for Beginners

Geometry FOR DUMMIES®

by Wendy Arnone, PhD

WILEY

Wiley Publishing, Inc.

Geometry For Dummies®

Published by
Wiley Publishing, Inc.
111 River Street
Hoboken, NJ 07030
www.wiley.com

Copyright © 2001 by Wiley Publishing, Inc., Indianapolis, Indiana

Published by Wiley Publishing, Inc., Indianapolis, Indiana

Published simultaneously in Canada

No part of this publication may be reproduced, stored in a retrieval system, or transmitted in any form or by any means, electronic, mechanical, photocopying, recording, scanning, or otherwise, except as permitted under Sections 107 or 108 of the 1976 United States Copyright Act, without either the prior written permission of the Publisher, or authorization through payment of the appropriate per-copy fee to the Copyright Clearance Center, 222 Rosewood Drive, Danvers, MA 01923, 978-750-8400, fax 978-646-8700. Requests to the Publisher for permission should be addressed to the Legal Department, Wiley Publishing, Inc., 10475 Crosspoint Blvd., Indianapolis, IN 46256, 317-572-3447, fax 317-572-4447, or e-mail permcoordinator@wiley.com

Trademarks: Wiley, the Wiley Publishing logo, For Dummies, the Dummies Man logo, A Reference for the Rest of Us!, The Dummies Way, Dummies Daily, The Fun and Easy Way, Dummies.com, and related trade dress are trademarks or registered trademarks of John Wiley & Sons, Inc. and/or its affiliates in the United States and other countries and may not be used without written permission. All other trademarks are the property of their respective owners. Wiley Publishing, Inc., is not associated with any product or vendor mentioned in this book.

For general information on our other products and services or to obtain technical support, please contact our Customer Care Department within the U.S. at 800-762-2974, outside the U.S. at 317-572-3993, or fax 317-572-4002.

Wiley also publishes its books in a variety of electronic formats. Some content that appears in print may not be available in electronic books.

Library of Congress Cataloging-in-Publication Data:

Library of Congress Control Number: 2001090690

ISBN: 0-7645-5324-0

Manufactured in the United States of America

10 9 8 7 6

3B/RV/QR/QU/IN

About the Author

Wendy Arnone has a PhD in Psychological Development from New York University's School of Education. She has a diverse background that touches on many areas of education. She has been published in several professional journals, and she's also a writer of children's books. She has worked as a content developer for Children's Television Workshop and the television network Noggin and as a consultant for renowned children's writer Mercer Mayer (of *Little Critter* fame). She has taught high school to adjudicated youth for the State of Connecticut and has taught college-level courses at New York University. She has served as an educational program evaluator for the Newark (New Jersey) Board of Education, as well as consulted for a private educational evaluation firm. As a researcher, she has received numerous awards. She was featured as a Dateline Discovery on *DatelineNBC.* She also has worked as the Assistant Director of Institutional Research at New York University's School of Dentistry. She has developed several educational titles — from concept to product completion — including an SAT, ACT, and PSAT preparatory CD-ROM and an elementary Spanish tutorial CD-ROM. And she is the president of her own consulting company that creates concepts and content for educational and corporate projects from written material to CD-ROMs to Web sites.

Author's Acknowledgments

I am grateful to the people at Wiley for their long hours and dedication to the production of this book. I would like to thank my development and copy editor, Sandy Blackthorn, for the long hours she spent improving the quality and content. Thanks to Alissa Schwipps, my project editor, for sticking with me through the seemingly endless review process. You have both been amazing. I appreciate the support of Susan Decker and Roxane Cerda for giving me the opportunity to write this book.

As always, I am grateful to my parents, family, and friends for their unwavering support. I wish I could list everyone individually but there simply isn't the space.

Special thanks to Suzanne Langlois, my aunt, for her original sketches, which the illustrations in this book are based on. And thanks to my sister Penny for being so supportive and my lil sis, Sarah, whom I hope will make good use of this book in her high school geometry class.

I appreciate the contributions of Ethan Blumenstrauch, Marshall Morrison, Jim Garrett, Andy Hunt, Debbie Eisenkraft, Kathy Gerber, Andrea Miller, Scott and Karalie Clark, Juleigh Walker, Cindy Ninos, Brady and Felicia Canfield, Wilhelmina Lydon, Leticia Fuentes Perez, James Lai, Frank Kubin, and Kevin Royer (my brother-in-law) — to this book and to my life.

And an extra special thanks to Trevor Christic for encouraging me to write this book. Everyone should have someone who believes in them.

Publisher's Acknowledgments

We're proud of this book; please send us your comments through our online registration form located at www.dummies.com/register.

Some of the people who helped bring this book to market include the following:

Acquisitions, Editorial, and Media Development

Project Editor: Alissa D. Schwipps

Acquisitions Editors: Susan Decker, Roxane Cerda

Copy Editor: Sandra Blackthorn

Technical Editor: David Herzog

Editorial Manager: Jennifer Ehrlich

Editorial Assistants: Carol Strickland, Jennifer Young

Cover Photo: Corbis Images/PictureQuest

Production

Project Coordinator: Maridee Ennis

Layout and Graphics: Karl Brandt, Jill Piscitelli, Betty Schulte, Rashell Smith, Erin Zeltner

Proofreaders: Melissa D. Buddendeck, David Faust, Susan Moritz, Marianne Santy, Charles Spencer

Indexer: Johnna VanHoose

Publishing and Editorial for Consumer Dummies

Diane Graves Steele, Vice President and Publisher, Consumer Dummies
Joyce Pepple, Acquisitions Director, Consumer Dummies
Kristin A. Cocks, Product Development Director, Consumer Dummies
Michael Spring, Vice President and Publisher, Travel
Brice Gosnell, Associate Publisher, Travel
Suzanne Jannetta, Editorial Director, Travel

Publishing for Technology Dummies

Richard Swadley, Vice President and Executive Group Publisher
Andy Cummings, Vice President and Publisher

Composition Services

Gerry Fahey, Vice President of Production Services
Debbie Stailey, Director of Composition Services

Contents at a Glance

Cartoons at a Glance

By Rich Tennant

"We all know it's a pie, Helen. There's no need to pipe the equation, 3.1415926853... on the top."

page 303

"The crew was wondering if there wouldn't be some sort of geometric way of proving the world is round instead of sailing up to the edge and hoping we don't fall off screaming into an endless black hole."

page 39

"I hear you think you got all the angles figured. Well, maybe you do and maybe you don't. Maybe the ratios of the lengths of corresponding sides of an equiangular right-angled triangle are equal, then again maybe they're not—let's see your equations."

page 193

You call this accurate course plotting? Now you know why they never ask you to draw crop circles anymore.

page 243

BASEBALL TRAPEZOID

STEALING 2ND
$A = \frac{1}{2}(b_1 + b_2) \times h$

"For the next month, instead of practicing on a baseball diamond, we'll be practicing on a baseball trapezoid. At least until everyone passes the geometry section of the GRE test."

page 65

Beyond Euclidean and Cartesian geometry, there is Ed Dubrowski geometry which proves that the volume of a sphere changes in proportion to the amount of food at an All-U-Can-Eat buffet.

page 313

$(xA - xB)^2 + (yA - yB)^2$

"...and so, with this formula, Descartes was able to compute the distance in a plane. However, despite this accomplishment, his luggage was still as likely to get lost on route as it had before."

page 7

Cartoon Information:
Fax: 978-546-7747
E-Mail: richtennant@the5thwave.com
World Wide Web: www.the5thwave.com

Table of Contents

Introduction

● ●

So you want to get to know geometry. This is the book for you. Geometry is a giant puzzle with interlocking pieces that you must snap together. Each piece fits snugly into another as you build, using simple steps, toward increasingly complex situations. Granted, these puzzle pieces don't have pictures on them. Instead, they contain basic snippets of knowledge that I explain to you in this book — in plain English and without a lot of technical details that would just bog you down.

I know the idea of figuring out geometry probably sounds a bit overwhelming right now, but it's really not. Figuring out geometry is like learning how to ride a bike. You start slow, with training wheels, and then before you know it, you're zipping down the road at unbelievable speeds. Same thing here. With this book by your side, before you know it, you'll have enough information to see how all the pieces fit together. So put on your geometry helmet and get ready to roll.

About This Book

This book comes with a warning — a good one, though, if there is such a thing. Whether you're in it for the long haul and plan to read it cover to cover, or you're just dabbling in a topic here and a topic there, this is not your average presentation of geometry. It's the Dummies version. Here, I bend the rules that are so staunchly adhered to in those *other* books. I do try to mention when I take liberties with the material so that if you see something presented differently somewhere else, you won't be confused. I don't compromise the accuracy of the material itself; I just present it a little differently. It's more casual.

Geometry For Dummies is for those of you who need a geometry boost or just want to engage in some fascinating reading. I can't do the work for you, but I can show you how to make it easier. The contents of this book are designed to give you a solid introduction to areas touched on by geometry — areas like these, for example:

Terms related to geometry

Measuring and making angles

Playing the proofing game

Relating to ratios

If you don't need info on all the topics covered, feel free to skip around. The detailed Table of Contents is the best place to start. If you need info on squares, then go to the quads chapter. No need to bother with circles. It isn't even necessary to read an entire chapter if you know what you're looking for. The chapters are broken into sections, and the Table of Contents can lead you right where you need to be.

Conventions Used in This Book

The following conventions are used throughout the text to make things consistent and easy to understand:

- Variables are in *italics* so that you don't get variable *a* (which represents, say, the apothem of a polygon) confused with the regular old letter a.

- Points (on a line or in a figure) are in *italics* too.

- New terms appear in *italics* with a definition in close proximity. You can also find words that appear in italics in the glossary — just in case.

- **Bold** is used to highlight the action parts of numbered steps (the actions you actually need to take).

- Certain statements in geometry are known as *postulates* and *theorems*. I introduce these statements in Chapter 1. They form the backbone of the reasoning in geometry. For reference purposes, I numbered the postulates and theorems (and — bonus! — corollaries) sequentially in each chapter. Any numbering of postulates, theorems, and corollaries is arbitrary (except if you're Euclid, the geometry god), so my numbering differs from any other book you pick up on geometry. Appendix C has a list of all the postulates, theorems, and corollaries in this book for quick reference.

- This book contains lots of proofs. They're two-column tables that prove a geometry concept. One column is a "Statements" column, and the other is a "Reasons" column. In geometry, you're not allowed to say anything without a good reason. Doing proofs ensures that a good trial of logic is followed. Everything must have evidence (or a reason). I go into detail on how a proof is assembled in Chapter 4.

- Concepts in geometry can be expressed in words, as in a sentence. Or they can be expressed in something I call geometric shorthand. Geometric shorthand uses symbols (like <) instead of words (like "less than"). I use a combo of both methods in this book — for you to get exposure to both methods. (Little tip of the day here: It takes less space to express things in geometric shorthand, so using shorthand is great on the "Statements" side of a proof.)

Besides using all the preceding conventions, throughout the text I include figures to help clarify the material being presented. Figures can do wonders for your understanding. They turn the abstract into the concrete. Nothing more solid than that.

Foolish Assumptions

I always try *not* to make assumptions. But due to page constraints, I have to. So, in this book, I make several assumptions about you, the reader:

- ✔ You can read, condense, and comprehend material. That's important.
- ✔ You can reason through a large pile of material. It's those proofs. You're not going to be able to make it through any discussion in geometry without the magical powers of deductive reasoning.
- ✔ You have a basic idea of what geometry is. I'm not talking about having all those postulates and theorems on the tip of your tongue. Just that you know that geometry deals with lines and angles and stuff.

Ya see? Not too much to ask.

How This Book Is Organized

This book is broken down, or divvied up, into seven parts. I guess how you look at it depends on your mood. (I hate downers, so I think it's divvied up.) Each of the seven parts addresses captivating material. Plus, because geometry has its own lingo, I definitely make sure that throughout the course of the book, we're speaking the same language. This section provides a description of what to expect in each part. I do, however, save some surprises for each chapter. It keeps things exciting.

Part 1: Just the Fundamentals (or Fun without Da Mental Drain)

Oh, yeah. This part's just what it says it is: fun, fun, fun. You get all cozy with the terms of geometry. And lines and angles become your best friends.

Part II: Getting the Proof

No claim can be deemed true without proof. Sometimes what you're looking for can be right there under your nose. But you have to know what you're looking for and know how to use it. After making your way through Part II, you can flaunt your powers of reasoning.

Part III: It Takes All Shapes and Sizes

The typical thoughts of shapes and sizes that come to mind when you think of geometry, it's all in here. I got yer triangles, quads, circles, and a bunch of stuff on polygons.

Part IV: Separate but Equal? (Inequalities and Similarities)

Sometimes I talk too much. In this part, I get to do just that — gab, gab, gab. I talk about theorems and postulates of inequality and similarity. I think that after reading this part, you too won't be able to resist the urge to talk up a storm about such things.

Part V: Geometry's Odds and Ends

This is where I clumped all the other stuff you should know that doesn't really fit in any of the other parts. It's a *smorgasbord* of geometry. Part V runs the gamut on topics. There's coordinates and working with the coordinate plane. It's actually kind of cool to work against something concrete. Locus is cool, too. You get to work with lots of points — points that, of course, follow rules. Trigonometry uses right triangles to gain information about things that may not be directly measurable. And last but not least, geometry with a hard body. Solids, that is. Three-dimensional shapes have invaded the world!

Part VI: The Part of Tens

This part contains lists: two to be exact. The first is a list of ten careers in case you're looking for opportunities to use the information from this book for the rest of your life. The second list contains ten hot tips to make your life with geometry easier.

Part VII: Appendixes

This part contains a useful mish-mash of information. You get some reference tables for squares and trigonometric functions; a bunch of formulae; and a list of all the postulates, theorems, and corollaries in this book. Oh, and I also tacked a glossary onto the end for good measure. It's hard to figure out what's going on when you don't know what the words mean.

Icons Used in This Book

In the margins of this book, you find some helpful icons that can make your journey easier:

Text marked with this bull's-eye icon gives helpful geometry pointers and information.

This icon sits next to information that you shouldn't forget. Take special note of this stuff because it pops up in multiple areas of geometry.

You need to be cautious not to commit the *faux pas* (that's hoity-toity French for *mistakes*) outlined by this icon.

Where to Go from Here

I know it seems like a lot, this geometry thing. I'm not going to lie to you — it is. Use this book as a reference or read it like a good novel, cover to cover. Whatever works best for you. If you get lost, then go back aways and read through the basics. The secret to making this stuff fun and manageable is the approach. And, baby, I more than got that covered.

Part I
Just the Fundamentals (or Fun without Da Mental Drain)

The 5th Wave   By Rich Tennant

©RICHTENNANT

$$\left(xA-{}_2B\right)^2 + \left(yA-yB\right)^2$$

"...and so, with this formula, Descartes was able to compute the distance in a plane. However, despite this accomplishment, his luggage was still as likely to get lost on route as it had before."

In this part . . .

Fun? Yes, you're here to have fun. Geometry *can* be fun. It's like putting a puzzle together. But a firm base is a necessity. If you build a house on a shaky foundation, the first time the wind blows the building collapses. I don't want to be pulling you out of any rubble.

The stuff in this part provides you with the terms involved in geometry and a description of what will become the objects of your affection — lines and angles. So read up! Get this stuff under your belt, and you'll be able to weather any storm.

Chapter 1

Getting Down to the Terms of Geometry

*Y*ou know that geometry is a math thing. That much you've got nailed down. But what you *don't* know is what geometry is exactly — or what kinds of things are involved with it. Well, you're at the right place. This chapter cuts to the chase with the basics. It explains the concept of geometry and defines the various thingamabobs that are used with it, plain and simple.

I just want to let you know that this is not your typical "Penguin in a tuxedo" totally stiff presentation of geometry. So loosen that bow tie and put on the cutoffs. Formal attire just isn't going to be permitted. Because...it's casual from here on out!

So What Exactly Is Geometry?

Well, how about the literal definition first: Geometry's origins come from the Greek word *geōmetria*. *Gē* means "earth," and *metre* means "measure." So, if we're talking literally here, *geometry* means "earth measure."

That aside, here's a doozie of a real-world definition, highbrow though it is: *Ordinary plane geometry* generally deals with the application of definitions, postulates, and theorems and is based on Euclid's work, *Elements,* from about 300 B.C.

And here, finally, is what you really need: In a nutshell, *geometry* is a section of math that involves the measurements, properties, and relationships of all shapes and sizes of things — from the tiniest triangle to the largest circle to the rectangle, and much more.

Euclid: The father of geometry

Euclid was a Greek mathematician who lived around 300 B.C. The exact dates of his life aren't known, but his bounty of work surely is. Euclid's best-known work is *Stoicheia,* which is Greek for "elements." In the twelfth century, Euclid's *Elements* was translated into Latin and took on the title *Elementa.* By whatever name, the work still marks the cornerstone of traditional geometry. Euclid's *Elements* contains 13 books and outlines postulates, theorems, and definitions for use within proofs. Two additional books, Books 14 and 15, are usually included in the text, but they aren't authored by Euclid. These books weren't part of his original work; they were added at a later point.

The following books from *Elements* are of particular interest to the development of geometry. You'll see the parallel as you explore the chapters of this book.

Book 1 contains info on triangles, including their construction and properties and the relation of their sides and angles to each other.

Book 3 contains the elementary geometry of the circle, including info on chords, secants, and tangents.

Book 4 explores problems resulting from inscribing polygons within circles and circumscribing polygons about circles. In particular, triangles and regular polygons are addressed.

Book 5 presents proportions and ratios, the basis for similar triangles.

Book 6 applies the theory of proportion from Book 5 to plane geometry. The info in this book was introduced by Pythagoras but tweaked by Euclid.

Books 11 through 13 deal with solid geometry.

Terms Related to Geometry

This section defines the various terms that are involved with geometry. Well, wait. I need to modify that. Because geometry involves some things called *undefined terms,* this section defines various terms involved with geometry *and* describes other terms that are pretty much undefinable.

But before digging into my definitions and (real word here, sure) descriptions themselves, you need to know what I consider to be the qualities of a good definition. There are four qualities as I see it, and I can dissect them by using the definition of a household cat as an example. Check out Angus in Figure 1-1.

Figure 1-1:
Angus is
just a cat
as per
definition,
just don't
tell him that!

The definition of a household cat: A *cat* is any animal of the class *Felis domestica* with retractile claws that kill mice.

My four qualities of a good definition

- The term being defined is named within the definition.

  In the definition of a household cat, I mention what I am defining — a cat.
- The term being defined is placed into a set or category.

  In the definition, I mention that a cat is of the class *Felis domestica*.
- The term being defined is differentiated from other terms without providing unnecessary information.

  In my definition, I mention that a cat has retractile claws that kill mice. Maybe the information "that kill mice" is more than you need to know. But that info may be necessary because it allows further differentiation of a cat from other animals.

✔ The definition is reversible.

Does this one still hold up? Reverse the definition and see: An animal of the class *Felis domestica* with retractile claws and kills mice. Sounds like a cat to me.

Terms so basic they can only be described

Geometry uses lots of defined terms, but many of those defined terms make use of undefined terms in their definitions. That may sound perplexing, but it's really not a big deal. Basically, *undefined terms* are words that are already so basic that they can't be defined in simpler terms, so they're described instead of defined. Undefined terms include a point, a line, and a plane.

A point

A *point* is represented by a dot, like a period on a page (see Figure 1-2). You name it by using a single uppercase letter. A point has no size and no dimension. Plainly put, that means it has no width, no length, and no depth. It only indicates a definite location or position. Essentially, other than indicating a location, a point has no physical existence.

Figure 1-2:
A point.

A.

A line

What's the quickest way to get from one place to another? A straight line. Yes, a concept of geometry can actually help you get to class on time. A *line* is straight and has no thickness (see Figure 1-3), and it's made up of a set of points that extends infinitely in both directions. The points that make up the line are called collinear points (see Figure 1-4). A line can be named by a lowercase letter, but, more commonly, it's named by any two points on the line.

Figure 1-3:
A line.

A ———•————————•———→ B

Figure 1-4:
Collinear
points,
whlch make
up a line.

X   Y   Z

A plane

No airports, no runways, no luggage. This plane doesn't fly. It only exists in two-dimensional (2-D) space, which means it has length and width but no depth. A *plane* in geometry is an infinite flat surface that has no boundaries and may be extended infinitely in any direction (see Figure 1-5). It is a set of all the lines that can be drawn through two intersecting lines. It is determined by exactly three non-collinear points. The flip-flop is also true; exactly one plane contains three non-collinear points (see Figure 1-6). A plane is indicated by a closed four-sided polygon and is named by a capital letter in one of its corners (as shown in Figure 1-5).

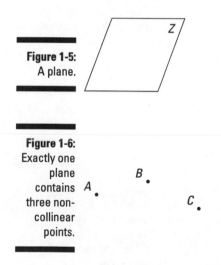

Figure 1-5:
A plane.

Figure 1-6:
Exactly one
plane
contains
three non-
collinear
points.

A   B   C

Terms that do have definitions

Defined terms in geometry can be defined (OK, yes, that's pretty intuitive) and abide by my four qualities of a good definition. Defined terms include a line segment, a ray, and an angle.

A line segment

A line segment, unlike a line, is not a never-ending story. It has a beginning, and it has an end. A *line segment* is a part of a line that has two endpoints that mark its finite length (see Figure 1-7). The names of these endpoints taken together are used to name the segment. Although the line segment may be identified by only two points, it is made up of not only those two endpoints but all the points between them. Because a line segment has a finite length, it can — unlike a line — be measured.

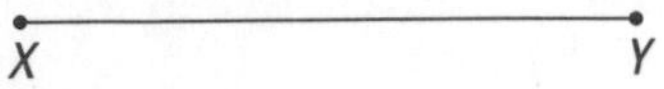

Figure 1-7:
A line
segment.

X _____ Y

A ray

A ray of sunshine begins at the sun and extends out into the sky. A ray has a beginning — it starts somewhere — but it has no end. A *ray* is a part of a line that has only one endpoint and extends infinitely in one direction (see Figure 1-8). Similar to a line segment, a ray has an infinite number of points on it. A ray is named by its endpoint and a point on the ray. In the letter pair that names the ray, the letter of the endpoint appears first.

Figure 1-8:
A ray.

A _____ B ⟶

An angle

If lines meet, they can form a relationship. In the social world of lines, this meeting is called an angle. The stereotypical angle looks something like the letter *V* (see Figure 1-9). An *angle* is the union of two rays or two line segments that meet at the point of the V. This point is called a common endpoint. The rays or line segments form the sides of the angle; the common endpoint, called the *vertex* of the angle (the vertex is the tip of the V), is used to name the angle. I just want to mention real quick that not all angles look like "V's." Which one's do and which don't are explored in Chapter 2.

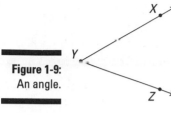

Figure 1-9:
An angle.

Postulates: A matter of trust

With life, it's easy to assume that something is true because it appears to be true. Same goes with geometry. In geometry, you can accept some statements or basic assumptions as being true without having to go to all the trouble of proving them. For those of you with trust issues, this concept may be a bit difficult to accept at first. But when you realize how much work it saves you, I'm sure you'll come around. The statements or basic assumptions that you can accept as being true are called *postulates.* Sometimes you see them referred to as *axioms.* And other times you may even see postulates and axioms separated out as two different kinds of statements. Regardless, the most important thing to remember about these statements is that you do *not* have to prove them. By learning these statements, you can save yourself a lot of time and aggravation when solving geometry problems.

Here's an example — your first postulate. It contains information about a line, and it's something that is self-evident but may actually be quite difficult to prove:

Postulate 1-1: Two points determine a line.

Translation? You need to identify two points in order to draw a straight line.

The word *postulate* actually comes from the Latin word *postulatus,* meaning "self-evident truth." The word *axiom* has its roots in the Latin word *axioma,* meaning "self-evident thing." I just know that explaining this kind of stuff will make the exploration of geometry just that much more fulfilling for you. Really.

Theorems: Prove it, babe

A theorem is in a way the opposite of a postulate. While a postulate is a statement accepted as true without proof, a *theorem* is a statement that you have to prove to be true. Postulates are actually used in the process of proving theorems. Proving a theorem is part of a process, one in which the next logical step is a geometric proof. I hesitate to say anything more here. (I go into all this proof stuff in Chapter 4.) I have included an example of a theorem below so that you can get a glimpse of all the fun to come!

Theorem 1-1: If two lines intersect, then they do so at exactly one point.

Closely related to the theorem is the corollary. It is a theorem that can be easily proved with the assistance of another theorem or postulate.

OK, that's a wrap. You're now acquainted with the basic terms you will encounter while studying geometry. It is time to put these terms to good use.

Chapter 2

Fishing Around with Lines and Angles

f you could follow a path to the stars, you could travel forever. Maybe you'd want to change your angle of approach at times or ascend at a slower rate, but the distance you could travel, it'd be unmeasurable.

But back down here on earth, things are more finite. How far do *you* need to go? Perhaps just to the local supermarket. With that kind of journey, you can measure the distance. How? With lines and angles. And that's the stuff this chapter's made up of.

How Lines Measure Up

So what particular line types of things can you measure? Is it possible to measure a line, a ray, or a line segment?

A *line* is forever. It extends infinitely in either direction, which makes it unmeasurable because it has no beginning and no end. A line is the ultimate never-ending story.

A *ray* is similar to a line, but it contains one endpoint. Its other end continues infinitely in the direction it's pointing to. Because a ray continues forever in one direction, it suffers the same fate as a line regarding measurability. Huh-uh — you can't measure a ray.

A *line segment,* though, has two endpoints. This distinction allows you to use a ruler to measure a line segment's finite length. Its length is the distance between its two endpoints.

Measuring a line segment

Line segments are measurable. So the decision at hand becomes how to measure a line segment and what unit of measure is most appropriate or meaningful, be it inches, centimeters, yards, or even kilometers. How to measure a line segment is covered in this section. How to determine the unit of measure is covered in your brain, which takes a look at just how big the line segment is and then makes a smart choice.

To measure a line segment, follow these steps:

1. **Place the end of a ruler on the center of one endpoint.**

2. **Determine the distance indicated on the ruler from the center of the first endpoint to the center of the second endpoint (see Figure 2-1).**

Figure 2-1:
Measuring
a line
segment
with a ruler.

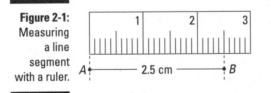

In Figure 2-1, the distance between point *A* and point *B* is 2.5 centimeters. Two ways are available to notate this result. The first way is to write the result like this: $m\overline{AB} = 2.5$. Obviously, just like a foreign language, geometric shorthand requires a bit of translation. The translation here is, "The measure of line segment $\overline{AB}$ equals 2.5." With the second method, you write the information like this: $\overline{AB} = 2.5$. The translation is, "The distance between points *A* and *B* equals 2.5." Both methods are correct and accurately convey the info. You can decide for yourself which method of notation you prefer to use. I use both methods in the examples throughout this book — just to keep you open-minded toward both methods.

Drawing a line segment

Just as with measuring a line segment, drawing one is pretty simple. It's an easy, four-step process:

1. **Put a dot on the paper where you want your first endpoint.**

2. **Get out your ruler and rotate it in the direction you want your line segment to go.**

3. **Measure how long you want your line segment to be and place a dot at the final point.**

4. **Join the two dots and make them more pronounced.**

Looking between the lines (points all over the place)

So I'm watching my favorite show on television, and a fast-food commercial comes on. All of a sudden, I *really* need a snack. I want to go from the couch to the refrigerator and back in record time — before the commercial's over. I need to determine the best possible route to take. I know that the shortest distance between any two points is a straight path, and luckily, it's a clear path from my couch to the fridge (see Figure 2-2). So off I go. But before I make it to the kitchen, I stop, think for a second (do I *really* want to ingest all those extra calories?), and then continue on my way (yes, I really, really do).

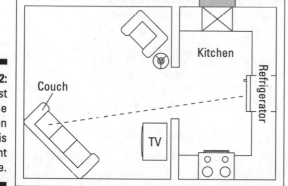

Figure 2-2:
The shortest distance between two points is a straight line.

My route to the kitchen can be represented as a line segment. The point at which I did my stopping and thinking was *between* the two endpoints (the couch and the fridge) that determined my route to the kitchen (see Figure 2-3). Although a line segment has two endpoints, it is also made up of all the points in between, even if they aren't named or identified.

Figure 2-3:
The two
endpoints
of a line
segment
and a point
in between.

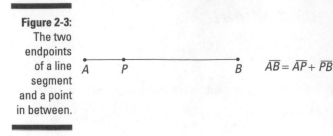

$\overline{AB} = \overline{AP} + \overline{PB}$

The Three Conditions of Between-Ness

For a point to be officially identified as being between the endpoints of a line segment, the following must occur (the following being what I call the Three Conditions of Between-Ness):

- ✔ The three or more points must be collinear.
- ✔ The points must have different locations on the line segment.
- ✔ When added together, the two segments created from the three points must equal the length of the larger (total) segment.

So to determine whether my stopping / thinking point was *officially* a point between endpoint couch and endpoint fridge, I need to see whether the preceding three criteria occurred. First, the three or more points on my route must be collinear. Take a look at Figure 2-3. Point *A* is the couch, point *B* is the refrigerator, and point *P* is where I stopped. All these points fall in a straight line. So the first quality of between-ness has been met. Next, the points must have different locations. Because I took steps between my positions at point *A,* point *P,* and point *B,* the three points do have different locations. Two down, one to go. Finally, I must determine whether the two segments of my route, when added together, equal the larger segment — or total distance of the route. The distance from point *A* to point *P* is 10 steps, the distance from point *P* to point *B* is 26 steps, and the distance from point *A* to point *B* is 36 steps. Yep, met that quality, too. $\overline{AB} = \overline{AP} + \overline{PB}$. So, as a result of meeting the Three Qualities of Between-Ness, I can conclude that I did stop at an official point between the endpoints of the couch and the refrigerator.

The Two Conditions of Midpoint-Ness

I'm going to rewrite history, just for a second. I've confirmed that I had stopped to think between the couch and the fridge. But what if I had stopped after 18 steps instead of 10? The representation of my path would look more like that in Figure 2-4. In Figure 2-4, not only is point *P* between points *A* and *B,* but it's also *halfway* between these points. A point that is located halfway from or exactly in the middle of the endpoints is called the *midpoint*. A midpoint divides a line segment into two equal, or *congruent*, segments.

A point is a midpoint of a line segment if

  ✔ The point is located between the endpoints.

  ✔ The two line segments created by point *P* are of equal length.

If I had stopped en route from the couch to the refrigerator at a point 18 steps from the couch, I would also be 18 steps from the fridge. In this scenario, both qualities would be satisfied ($\overline{AP} = \overline{PB}$), making point *P* the midpoint (see Figure 2-4).

Figure 2-4:
The midpoint of a line segment.

$$\overline{AP} = \overline{PB}$$

The Interchangeable-Ness Factor

Exercise can be a good thing, and I really hate being a couch potato. After my favorite TV show is over, I usually take a walk. I live in the middle of the block in my neighborhood and can go a half-block to my right or a half-block to my left. Either way is the same distance and provides the same amount of exercise, so it doesn't matter which way I choose to go. Either distance is interchangeable. The same is true for line segments of equal distance. They're interchangeable.

Theorem 2-1: Given a line segment $\overline{AD}$ with points *B* and *C* between endpoints *A* and *D*, if $\overline{AB} = \overline{CD}$, then $\overline{AC} = \overline{BD}$.

Translation: If two segments within a line segment are equal, then they can be interchanged. In Figure 2-5, for example, $\overline{AB} = \overline{CD}$, and they can be substituted for each other in any proof.

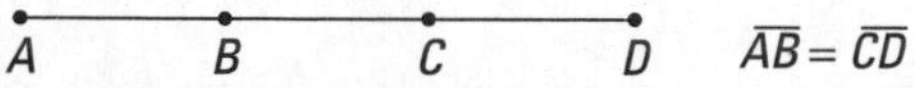

Figure 2-5:
Line segments of equal length are inter-changeable.

$$\overline{AB} = \overline{CD}$$

Looking at Angles from, Um, All Angles

On a map, you trace your route and come to a fork in the road. Two diverging roads split from a common point and form an *angle*. The point at which the roads diverge is the *vertex*. An angle separates the area around it, known in geometry as a plane, into two regions. The points inside the angle lie in the interior region of the angle, and the points outside the angle lie in the exterior region of the angle.

When two rays or line segments, for that matter, form an angle, the angle itself has a name. So how do you call an angle? Well, not with *Suey!* You'd look pretty silly making pig calls at your book. No, there are better ways to call, or name, an angle. Three of them. And you also need the symbol ∠ to represent an angle.

The three ways to name an angle are

- ✔ Use the capital letter of its vertex (see Figure 2-6).

- ✔ Use a lowercase letter or a number placed inside the vertex of the angle (see Figures 2-7a and 2-7b).

- ✔ Use three capital letters — one letter representing a point on each of the sides and another letter for the vertex (see Figure 2-8). The letter for the vertex is always in the middle.

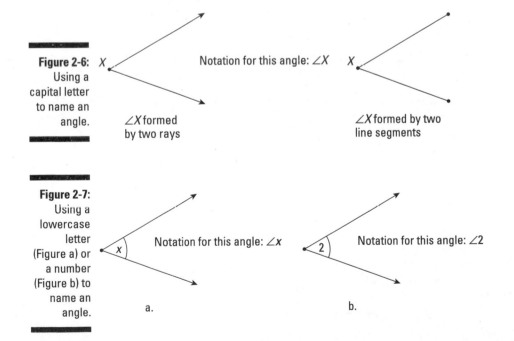

Figure 2-6:
Using a capital letter to name an angle.

Notation for this angle: ∠X

∠X formed by two rays

∠X formed by two line segments

Figure 2-7:
Using a lowercase letter (Figure a) or a number (Figure b) to name an angle.

Notation for this angle: ∠x

Notation for this angle: ∠2

a.

b.

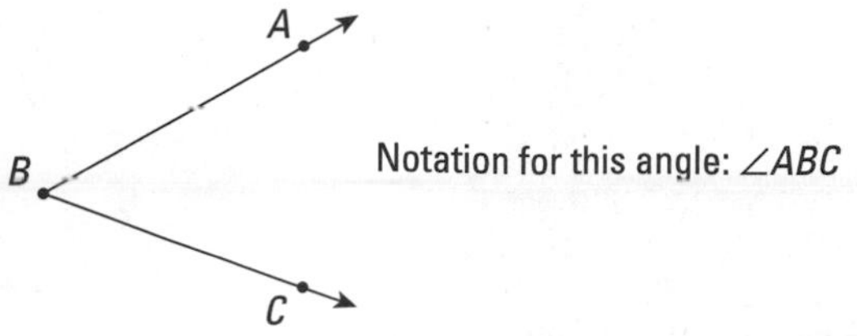

Figure 2-8:
Using three capital letters to name an angle.

Notation for this angle: $\angle ABC$

Bottom line? When you get a puppy, the first thing you do is name it. Think of an angle like a puppy and do the same thing with it.

The breeds of angles

If you think of an angle like a puppy, you name that angle right away. Then you need to figure out what breed of angle you have. Several different angle breeds, or types, exist. You can figure out what breed of angle you have by its measure. The most common measure of an angle is in degrees. (Yep, as in taking your temperature. Here, though, only the name is the same. This stuff has nothing to do with how hot or cold the angle is.) I show you how to take an angle's temperature (so to speak) in the next section. Here, I want to introduce you to the four types of angles:

✔ **Right angle.** With this angle, you can never go wrong. The right angle is one of the most easily recognizable angles. It's in the form of the letter L, and it makes a square corner (see Figure 2-9). It has a measure of 90 degrees.

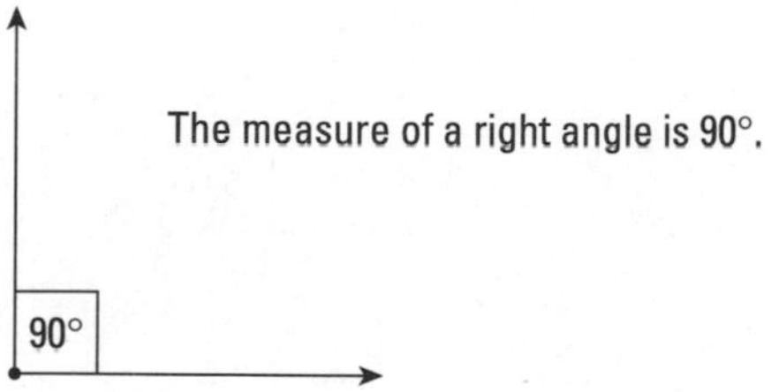

The measure of a right angle is 90°.

Figure 2-9:
A right angle.

90°

✔ **Straight angle.** You know what? It's actually a straight line. Most people don't even think of this type as an angle, but I assure you, it is. A straight angle is made up of opposite rays or line segments that have a common endpoint (see Figure 2-10). This angle has a measure of 180 degrees.

Right and straight angles are pretty easy to spot just by looking at them, but never jump to conclusions about the measure of an angle. Being cautious is best. If the info isn't written on the page, don't assume anything. *Measure.*

Figure 2-10:
A straight
angle.

The measure of a straight
angle is 180°.

180°

✔ **Acute angle.** It's the *adorable* angle.

Actually, it's just a pinch. It's any angle that measures more than 0 degrees but less than 90 degrees. An acute angle falls somewhere between nonexistent and a right angle (see Figure 2-11).

✔ **Obtuse angle.** This type is just not as exciting as an acute angle. It's measure is somewhere between a right angle and a straight angle (see Figure 2-12). It is a hill you must climb, a mountain for you to summit. It has a measure of more than 90 degrees but less than 180 degrees.

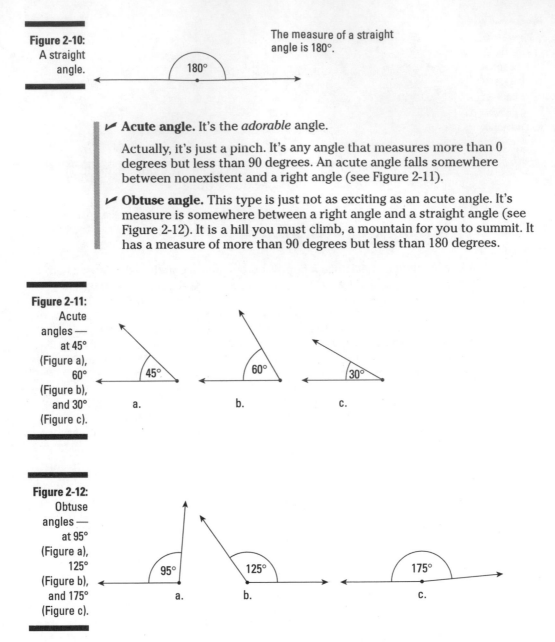

Figure 2-11:
Acute
angles —
at 45°
(Figure a),
60°
(Figure b),
and 30°
(Figure c).

45°
a.

60°
b.

30°
c.

Figure 2-12:
Obtuse
angles —
at 95°
(Figure a),
125°
(Figure b),
and 175°
(Figure c).

95°
a.

125°
b.

175°
c.

Measuring and making angles

This section shows you how to measure and make angles. To do both tasks, you use a protractor, a very useful tool to keep around (see Figure 2-13).

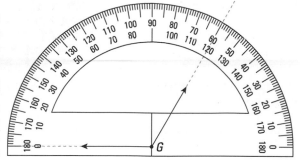

Figure 2-13:
The handy-
dandy
protractor,
for
measuring
and making
angles.

When choosing a protractor, I always like to get one made of clear plastic. Figuring out the measure of an angle is easier because you can see the line for the angle through the protractor.

Most protractors have two scales. One scale goes left to right. And the other scale, you guessed it, reads right to left. You need to make a choice as to which one to use. Decisions, decisions.

Measuring 'em

Angles are most commonly measured by degrees, but for those of you who are sticklers for accuracy, even smaller units of measure can be used: minutes and seconds. These kinds of minutes and seconds are like the ones on a clock — a minute is bigger than a second. So think of a degree like an hour, and you've got it down: One degree equals 60 minutes. One minute equals 60 seconds.

The relationship between these units of measure can also be written in geometric shorthand:

Relationship between Measures	Relationship in Geometric Shorthand
1 degree equals 60 minutes.	$1° = 60'$
1 minute equals 60 seconds.	$1' = 60''$

Now it's time to put your eyes to use. Before measuring an angle, the first thing you need to do is spec it out and *estimate* which type you think it is. Is it a right angle? A straight angle? Acute or obtuse? After you estimate it, then measure the angle. Follow these steps:

1. **Place the notch or center point of your protractor at the point where the sides of the angle meet (the vertex).**

2. **Place the protractor so that one of the lines of the angle you want to measure reads zero (that's actually 0°).**

   Using the zero line isn't necessary because you can measure an angle by getting the difference in the degree measures of one line to the other.

It's easier, however, to measure the angle when one side of it is on the zero line. Having one line on the zero line allows you to read the measurement directly off the protractor without having to do more math. (But if you're up for the challenge of doing more math, hey, knock yourself out.)

3. **Read the number off the protractor where the second side of the angle meets the protractor.**

Take some advice from someone who's been there:

✔ Make sure that your measure is close to your estimate. Doing so tells you whether you chose the proper scale. If you were expecting an acute angle measure but got a seriously obtuse measure, you need to rethink the scale you used. Try the other one.

For example, ∠G in Figure 2-13 should have a degree measure greater than 90° but less than 180° because I cite this angle as obtuse.

✔ If the sides of your angle don't reach the scale of your protractor, you should extend them so that they do. Doing so increases the accuracy of your measure.

For example, in Figure 2-13, the measure of ∠G is 120°. The sides of the angle come up a bit short and can't quite reach the protractor, so I extended 'em. I get a more accurate measure that way. It cuts down on the margin of error.

✔ Remember that the measure of an angle is always a positive number.

So what do you do if your angle doesn't quite fit on the protractor's scale? Look at Figure 2-14 for an example. The angle in this figure has a measure of greater than 180°. Now what? Sorry, but in this case, you're going to have to expend a little extra energy. Yes, you have to do some math. These angles are known as reflex angles and they have a measure of greater than 180°.

Figure 2-14:
Reflex angles don't fit on the protractor's scale, so you gotta do some math to measure 'em.

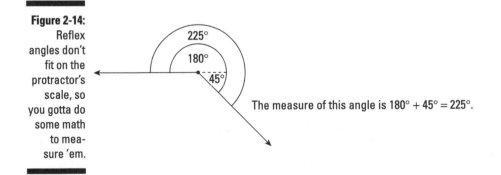

The measure of this angle is 180° + 45° = 225°.

Draw a line so that you have a straight line (see the extended dots on Figure 2-14). The measure of this portion of the angle is 180° because it's a straight angle. Now measure the angle that is formed by the extension line you just made and the second side of the original angle you want to measure. (If you get confused, just look at Figure 2-14.) Once you have the measure of the second angle, add that number to 180. The result is the total number of degrees of the angle. In Figure 2-14, 180° + 45° = 225°.

Making 'em

If you want to branch out and expand your horizons, you can actually draw your own angles. Doing so comes in handy for problems that don't have fig- ures with them. No longer will you be a slave to your circumstances.

To draw your own angle, follow these steps:

1. **Draw a ray (or line segment) and place the notch or center of the protractor on the endpoint of the ray.**

   For this example, call the ray $\overrightarrow{YZ}$.

2. **Move the protractor so that $\overrightarrow{YZ}$ is on the zero line.**

3. **Look along the scale of the protractor and find the number that repre- sents the measure of the angle you want to draw.**

   In this case, draw an angle of 60°.

4. **Place a dot under this number.**

   Call it point X.

5. **Draw a line connecting point Y and point X.**

   Point Y is the vertex of the angle.

You've just drawn $\angle XYZ$ (see Figure 2-15). The m$\angle XYZ$ is 120°.

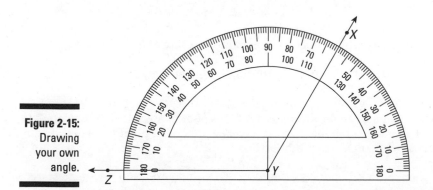

Figure 2-15:
Drawing
your own
angle.

Angles can have neighbors

An angle doesn't exist in a vacuum. It can have neighbors. A neighboring angle shares a common side and a common vertex with another angle. Collectively, these angles are commonly referred to as *adjacent angles* because they're next to each other on the same plane. Figure 2-16 shows an example: ∠ABX and ∠XBC are adjacent angles. Although they share a side and a vertex, these two angles don't share everything. They selfishly covet their own interior points and don't share them with their neighbors.

Figure 2-16:
Neighborly
critters —
adjacent
angles.

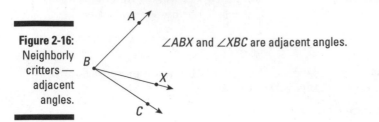

∠ABX and ∠XBC are adjacent angles.

With adjacent angles, the whole angle is the sum of its two smaller angles. That is, adjacent angles can be added and subtracted just like numbers (see Figure 2-17).

Figure 2-17:
Adding the
measures of
two angles
makes one
big angle.

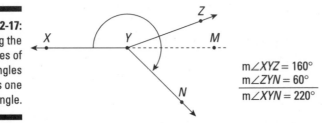

$$m\angle XYZ = 160°$$
$$m\angle ZYN = 60°$$
$$\overline{m\angle XYN = 220°}$$

Postulate 2-1: The whole quantity is equal to the sum of all of its parts.

Translation: Add all the parts together, and you get the whole quantity.

Postulate 2-2: If ray $\overrightarrow{YN}$ is interior to angle *XYZ*, then the measure of angle *XYN* and the measure of angle *NYZ* equals the measure of angle *XYZ*. Postulate 2-1 is kind of a general statement. When it is specifically applied to angles, it can be referred to as the Angle Addition Postulate.

Translation: m∠XYZ = m∠XYN + m∠NYZ. The Angle Addition Postulate states that if the measures of two adjacent angles are added together, the measure of the whole angle equals the measure of the two adjacent angles.

Take a look at Figure 2-18. If ∠*ABX* and ∠*CBX* have equal angle measures, then ray *XB* is referred to as an *angle bisector*. An angle bisector is a ray that has an endpoint at the bend of an angle and divides the angle fairly into two equal, or *congruent*, angles. Congruent angles are indicated by stroke marks — same angle measures, same number of stroke marks. In Figure 2-18, for example, m∠*ABX* = 23° and m∠*XBC* = 23°. So ∠*ABX* and ∠*XBC* are congruent.

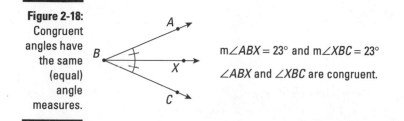

Figure 2-18:
Congruent angles have the same (equal) angle measures.

m∠*ABX* = 23° and m∠*XBC* = 23°

∠*ABX* and ∠*XBC* are congruent.

Angles can travel in pairs

Several different types of angles travel in pairs. These angles include complements, supplements, and vertical angles.

Complements

If I said that an angle looked marvelous, would that be a complement? A compliment, yes, but not a complement. A *complement* actually refers to pairs of angles whose measures add up to 90°. If two angles are complementary, and if I know the measure of one of the angles, I can find the measure of the other angle by using subtraction. I subtract the known measure of the angle from 90 and, presto, I have the measure of the second angle. Figure 2-19 shows complementary angles.

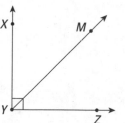

Figure 2-19:
The measures of complementary angles add up to 90°.

m∠*XYM* = 45° and m∠*MYZ* = 45°
m∠*XYM* + m∠*MYZ* = 90°

∠*XYM* and ∠*MYZ* are complementary.

Supplements

Any two angles whose measures equal 180° are *supplementary*, no vitamins needed. If supplementary angles are also adjacent angles, then the sides of the angles that are not shared form a straight line. Figure 2-20 shows supplementary angles.

Figure 2-20:
The measures of supplementary angles add up to 180°.

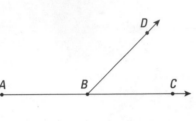

m∠*ABD* = 135° and m∠*DBC* = 45°
m∠*ABD* + m∠*DBC* = 180°

∠*ABD* and ∠*DBC* are supplementary.

A pair of angles doesn't need to be adjacent to be complementary or supplementary.

Verticals

Vertical angles sit across from each other but never next to each other. They are opposite angles and not adjacent angles. Vertical angles are formed when two lines traveling on converging paths finally collide and then continue on their way (see Figure 2-21). Vertical angles have the same degree measure.

Figure 2-21:
Vertical angles are opposite angles. They sit across from each other.

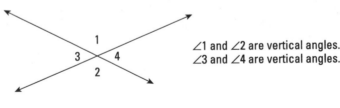

∠1 and ∠2 are vertical angles.
∠3 and ∠4 are vertical angles.

For good measure: Theorems for congruent angles

Congruent angles — those that have the same (equal) measures — have a slew of theorems attached to them. And some of these theorems revolve around the likes of (jargon alert here) complementariness and supplementariness and even verticalness.

Theorems 2-2 – 2-8: Two angles are congruent if any of the following are true:

✔ *Theorem 2-2:* They are both right angles.

✔ *Theorem 2-3:* They are both straight angles.

✔ *Theorem 2-4:* They are complements of the same angle or a congruent angle.

✔ *Theorem 2-5:* Their complements are congruent.

✔ *Theorem 2-6:* They are supplements of the same angle or a congruent angle.

✔ *Theorem 2-7:* Their supplements are congruent.

✔ *Theorem 2-8:* They are vertical angles.

Crossing over: Lines that form angles

The point of collision of two lines or line segments is known as the *point of intersection* (see Figure 2-22). The effect that the collision has on one line is determined by the approach of the other line. For example, in Figure 2-23, the effect that the collision has on $\overline{AB}$ is determined by the approach of $\overleftrightarrow{XY}$. If $\overleftrightarrow{XY}$ intersects $\overline{AB}$ at its midpoint, then $\overleftrightarrow{XY}$ bisects $\overline{AB}$. $\overleftrightarrow{XY}$ is then referred to as a *segment bisector.*

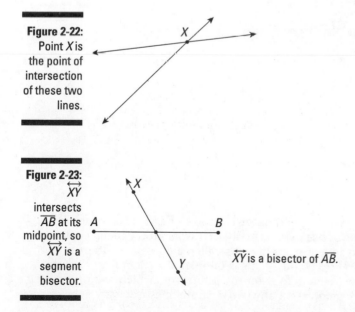

Figure 2-22: Point *X* is the point of intersection of these two lines.

Figure 2-23: $\overleftrightarrow{XY}$ intersects $\overline{AB}$ at its midpoint, so $\overleftrightarrow{XY}$ is a segment bisector.

$\overleftrightarrow{XY}$ is a bisector of $\overline{AB}$.

If the intersection of the two lines forms a right angle, then the lines are perpendicular. *Perpendicular* describes the relationship the lines have to each other. It has to do with the measure of the angles that are formed where the lines meet. The measure of the angles of perpendicular lines is always 90° (see Figure 2-24). As usual in geometry, a symbol is available to either complicate things or make them easier. I guess it depends on how you look at it. Anyway, the symbol for perpendicular lines is ⊥.

For perpendicular lines, picture in your head the hands of a clock when it's 3 o'clock or 9 o'clock.

Figure 2-24:
A slew of perpendicular lines.

Theorem 2-9: A perpendicular line segment is the shortest segment that can be drawn from a point to a line.

Translation: Look at Figure 2-25. $\overline{AB}$ and $\overline{AC}$ are both line segments drawn to $\overleftrightarrow{BD}$. $\overline{AC}$ is perpendicular to $\overleftrightarrow{BD}$. $\overline{AB}$ is not perpendicular to $\overleftrightarrow{BD}$. $\overline{AB}$ is longer than $\overline{AC}$.

Figure 2-25:
A perpendicular line segment is the shortest route from a point to a line.

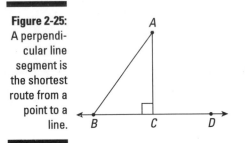

A line may come into contact with more than one line as it continues on its journey. In fact, it may cross over as many as it wants to. A special name is available for such an adventurous line. It's called a *transversal*. It crosses over two or more lines each and — this is important — at *different* points. A transversal has the freedom to enter the scene from whichever direction it wants to because it can intersect any two or more lines from any angle (see Figure 2-26).

Figure 2-26:
$\overleftrightarrow{MN}$ is a
transversal
line — one
that crosses
over two or
more lines
at different
points.

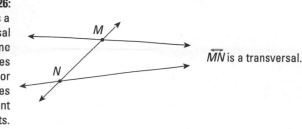

$\overline{MN}$ is a transversal.

Ships that pass in the night

Sometimes paths cross; sometimes they don't. If two ships are sailing to an island, and their routes don't intersect at any point, their routes are either parallel or skew in relationship to each other. If they're on parallel routes, the two ships will never be at the same exact location regardless of the time of day (see Figure 2-27). If Ship A is dragging a trunk of gold behind it and the trunk gets a hole in it, leaving a trail of gold, Ship B will never cross over the path of gold. And Ship B will never be able to recover the gold from the sea because it'll never travel over any of Ship A's route. Unless, of course, the treasure drifts. But it's gold, so it won't. It'll sink.

Figure 2-27:
On parallel
routes, two
ships will
never be at
the same
exact
location.

Routes that are parallel are indicated by the symbol ∥. If this symbol doesn't appear in the information given, don't assume that the pair of lines is parallel. Parallel lines can be horizontal, vertical, or slanted. The direction they travel isn't what makes lines parallel; what makes them parallel is their relationship to each other. Parallel lines are usually indicated with arrowheads (see Figure 2-28).

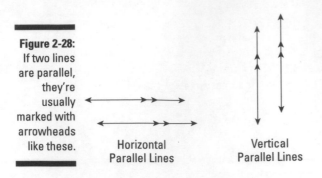

Figure 2-28:
If two lines
are parallel,
they're
usually
marked with
arrowheads
like these.

Horizontal
Parallel Lines

Vertical
Parallel Lines

Postulate 2-3: Given a line and a point not on that line, exactly one parallel line may be drawn through the given point.

This one's the Parallel Postulate — sometimes referred to as Euclid's Fifth Postulate because it's based on the fifth postulate listed in Euclid's *Elements*.

Translation: If you have a line, and you have a point not on that line, only one line can be drawn through the point that is parallel to the existing line (see Figure 2-29).

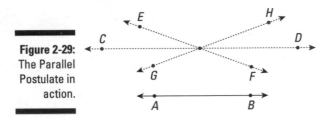

Figure 2-29:
The Parallel
Postulate in
action.

In Figure 2-29, the only line that is parallel to $\overleftrightarrow{AB}$ is $\overleftrightarrow{CD}$. $\overleftrightarrow{EF}$ and $\overleftrightarrow{GH}$ will eventually intersect with $\overleftrightarrow{AB}$ if they are extended.

Postulate 2-4: Two lines in the same plane either run parallel to each other or intersect.

Translation: Lines in the same plane either intersect and form angles or run parallel to each other and don't form any angles.

Ships that transverse in the night

Two ships are traveling along, and a third ship sails behind them crossing both their paths. The route of the third ship is a transversal in relationship to the routes of the other two ships. A *transversal* crosses two or more lines at different points.

When lines meet, they form angles. Two intersecting lines form four angles: two adjacent angles and two vertical angles. When two lines are intersected by a transversal, eight angles are created. Four angles are formed for each line the transversal crosses (see Figure 2-30). If the transversal crosses 3 lines, 12 angles are formed.

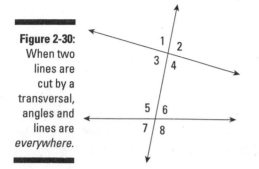

Figure 2-30: When two lines are cut by a transversal, angles and lines are *everywhere.*

In addition to interior and exterior angles, three types of angle pairs are also formed: alternate interior, corresponding, and alternate exterior angles. Here's a quick reference for transversal lingo (think of it as orientation):

- ✔ **Interior:** Between parallel lines.
- ✔ **Exterior:** Outside parallel lines.
- ✔ **Corresponding:** In the same relative location.
- ✔ **Alternate:** Opposite sides of the transversal.

Alternate interior angles are interior angles on either side of the transversal. They form a Z shape (see Figure 2-31). *Corresponding angles* are on the same side of the transversal and in the same position relative to the lines that the transversal crosses. They form an F shape (see Figure 2-32).

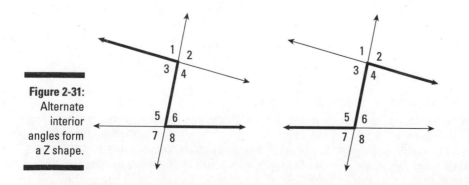

Figure 2-31: Alternate interior angles form a Z shape.

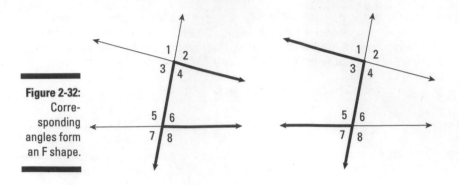

Figure 2-32:
Corre-
sponding
angles form
an F shape.

The world isn't always orderly. The transversal can be standing straight up, tipped, tilted, or slanted. You get the picture. The backbone of the Z shape and the F shape always falls on the transversal. So trace along the transversal to make it a heavier line. That way, you'll have a better view of which angles fall where.

Sorry — alternate exterior angles don't form a shape. You're going to have to find these angles by using brainpower. Just put together what you know, and you'll find them. All exterior angles are outside the lines that are transversed. *Outside* means not between the lines that the transversal crosses. *Alternate* means opposite sides of the transversal. I think a peek at Figure 2-33 is proba-bly the best course of action.

Figure 2-33:
Alternate
exterior
angles are
outside the
transversed
lines and
are on
opposite
sides of the
transversal.

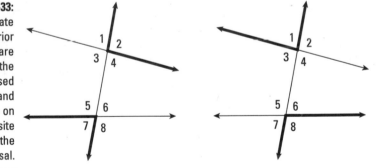

Transversing the parallel

When a transversal crosses two parallel lines, you have a special occasion (see Figure 2-34). It's time to celebrate the additional information about which angles are congruent or which angles are supplementary to each other. Trust me, this is a big deal, especially when you're trying to determine an angle's measure and you don't have a protractor.

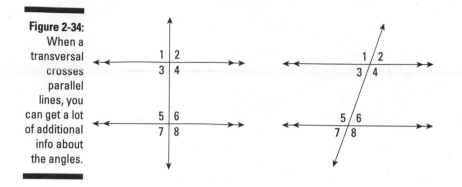

Figure 2-34:
When a transversal crosses parallel lines, you can get a lot of additional info about the angles.

Postulate 2-5: If two lines are crossed by a transversal, and the alternate interior angles are congruent, then the lines are parallel.

Translation: If two lines are crossed by a transversal, and you know that its alternate interior angles are equal, you can conclude that the lines are parallel.

The flip-flop, or converse, is also true:

Postulate 2-6: If two parallel lines are crossed by a transversal, then their alternate interior angles are congruent.

Translation: If you have some parallel lines, then their alternate interior angles have the same measure.

Postulate 2-7: If two lines are perpendicular to a transversal, then these two lines are parallel to each other.

Translation: If you have two lines that are perpendicular to a transversal, both of these lines hit the transversal at the same angle (90°), making the two lines parallel to each other.

OK, I've got one more angle pair for you. Although this angle pair exists with any transversal, it only gives you useful information if the transversal crosses parallel lines. These angles are called *same-side interior angles*. They're interior angles that are on the same side of the transversal. The sides of these two angles form a U shape (see Figure 2-35).

Want some theorems about parallel lines? You got it:

- ✔ *Theorem 2-10:* If a pair of corresponding angles is congruent, then the lines are parallel.
- ✔ *Theorem 2-11:* If a pair of alternate exterior angles is congruent, then the lines are parallel.
- ✔ *Theorem 2-12:* If a pair of same-side interior angles is supplementary, then the lines are parallel.

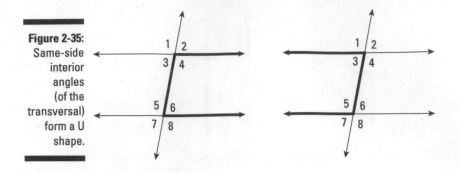

A flip side, or converse, of Theorems 2-10 through 2-12 exists, too. The following theorems help if you start off knowing that the lines are parallel and need to show something about the angles:

- ✓ *Theorem 2-13:* If lines are parallel, then any pair of corresponding angles is congruent.

- ✓ *Theorem 2-14:* If lines are parallel, then any pair of alternate exterior angles is congruent.

- ✓ *Theorem 2-15:* If lines are parallel, then any pair of same-side interior angles is supplementary.

The bottom line

Here's the big, general rule about angles and transversals:

Theorem 2-16: If two parallel lines are cut by a transversal, any pair of angles is either congruent or supplementary.

Translation: If two parallel lines are crossed by another line, pick any two angles and they'll either be congruent or supplementary (*supplementary* meaning that the sum measure of their angles equals 180°).

Want some examples? Use the angles in Figure 2-34 to view these angle relationships:

Angle Pair	*Relationship to Each Other*
$\angle 1$ and $\angle 2$	Supplementary
$\angle 1$ and $\angle 3$	Supplementary
$\angle 1$ and $\angle 4$	Congruent (vertical angles)
$\angle 1$ and $\angle 5$	Congruent (corresponding angles)
$\angle 1$ and $\angle 6$	Supplementary ($\angle 6 = \angle 2$)
$\angle 1$ and $\angle 7$	Supplementary ($\angle 7 = \angle 3$)
$\angle 1$ and $\angle 8$	Congruent ($\angle 8 = \angle 4$)

Part II
Getting the Proof

The 5th Wave — By Rich Tennant

"The crew was wondering if there wouldn't be some sort of geometric way of proving the world is round instead of sailing up to the edge and hoping we don't fall off screaming into an endless black hole."

In this part . . .

The accused is the theorem. You are the plaintiff. The burden of proof is on you to show that a theorem is true. You can't just make a statement without having anything to back it up. You can't just throw out a bunch of potentially false haphazard statements. So break out your funny inspector's hat and magnifying glass. You need to look for clues and piece them together.

In this part, you get the discipline and direction you need to discover the clues that will lead you from the given information to the solution. Things fall into place if you just let them come. Trust me.

Chapter 3

Logically Speaking about Proofs

I love a good mystery. It's like a puzzle. You have to put the pieces together until they form a picture that makes sense. Same goes with proving things in geometry. You have to put the pieces of what you're trying to prove together until they form a picture that makes sense.

But the first thing you've got to do is figure out how to put the pieces together. And you have to know how to justify statements used in proofs. That's what this chapter's all about. After you read the stuff in this chapter, you'll have the information you need to form an approach and some methods to get from the given information to a conclusion.

Methods of Reasoning from the Sherlock Holmes Handbook

Sherlock Holmes was always able to solve a case. I hope that your record of solving geometry cases will be just as good. But how do you do it?

Well, before you can go about doing it, you first must understand the process involved in solving a case: You've got to think before you act. (How many times have you heard *that*?) The reasoning process can take several forms. It can be intuitive, inductive, deductive, or indirect.

Intuitive reasoning: The lightbulb moment

Intuition is that "I just know it" feeling. It's not really based on anything. You jump to a conclusion without a thorough analysis of the facts. This method isn't very accurate because it doesn't follow any formal reasoning or require any evidence. What you may save in time you may lose in accuracy. So look before you leap.

Inductive reasoning: Under observation

With inductive reasoning, you run less of a risk of making an error than with intuition. Unlike intuition, inductive reasoning actually involves an organized attempt to test a theory. But it still leaves room for bias because it involves a process of drawing conclusions from limited observation and testing.

Inductive reasoning follows a six-step process:

1. **Carefully state the problem so that no available information is omitted.**

2. **Create known examples and test conditions for the information.**

3. **Make measurements and observations.**

4. **Observe a pattern in the data.**

5. **Assume that the observed pattern won't end with further tests.**

6. **Draw a conclusion from the pattern.**

The data used in inductive reasoning is based on experimentation and observation and may not accurately reflect what's really going on. You know what they say: "Garbage in, garbage out." A conclusion based on faulty testing and observation may, itself, be faulty.

With inductive reasoning, you reach a conclusion based on something that has a high degree of probability for occurrence. But what happens if, during the process of testing and experimentation, you find even *one* instance that doesn't follow the data pattern that your conclusion was drawn from? You have to alter your conclusion to include the new information, or worse — throw out your conclusion and start all over again.

Deductive reasoning: Just the facts

Now here's a method of reasoning in which conclusions are drawn only after a step-by-step process that's based on facts. This type of reasoning is a funneling process: You take all that is given and, bit by bit, provide statements of proof to arrive at a conclusion.

You can prove the truth of a theory in three steps:

✔ Make a general statement about a whole group. This statement is known as a *major premise*.

✔ Make a specific statement about something that indicates membership in the group. This statement is known as a *minor premise*.

If you accept both the major and minor premises as true, then you must accept the conclusion as true.

✔ Make a statement (your conclusion) indicating that the information applies to the group (the major premise) and to the individual (the minor premise).

Examples would be helpful, huh? No problem. Read on.

The cat's meow: A drawing-a-valid-conclusion example

Start with a general statement about a whole group — your major premise:

All cats have whiskers.

Now make a specific statement about something that indicates membership to the group — your minor premise:

Domestic shorthairs are cats.

So you've indicated that domestic shorthairs are members of the group *cats*.

Now you can arrive at a conclusion based on the major and minor premises:

Domestic shorthairs have whiskers.

This series of statements is called a *syllogism*. You arrive at your conclusion based on statements accepted as true. If the major and the minor premises are true, then the conclusion drawn from these premises must also be true.

But as is usual in life, this isn't always the case. You can still draw invalid conclusions even if your premises are true.

Does he or doesn't he? A drawing-an-invalid conclusion example

This is a story about Kevin, and he lives in Connecticut where the age to obtain a driver's license is 16 years. Kevin is over the age of 16. Here's the information as a syllogism:

✔ **Major premise:** A person must be at least 16 years old to get a driver's license in Connecticut.

✔ **Minor premise:** Kevin is over the age of 16.

✔ **Conclusion:** Kevin has a driver's license.

As you can see, both the major and the minor premises are true, but the conclusion is lacking. The relationship between the groups must make sense in order for the conclusion to be valid. Here, the progression doesn't make sense. Just because Kevin is over the age of 16 doesn't mean he has a driver's license. Yeah, he possibly does, but from the information given, there's not conclusive evidence that he does.

Let me alter the story about Kevin a little bit. If I know he lives in Connecticut where the age to obtain a driver's license is 16 years and I also know that he has a valid driver's license, then I can make some sense of this info. Now check out the sequence of statements. Doesn't it make more sense?

- ✔ **Major premise:** A person must be at least 16 years old to get a driver's license in Connecticut.
- ✔ **Minor premise:** Kevin has a driver's license.
- ✔ **Conclusion:** Kevin is over the age of 16.

If Kevin must be at least 16 years old to hold a valid driver's license, and he has a driver's license, then it follows from the information that Kevin is over the age of 16.

Be careful not to draw invalid conclusions from data.

"Hey, is there any actual geometry in this chapter?"

Of course there is. Right now, in fact. You know about right angles (see Figure 3-1), so you can apply some deductive reasoning to them. (If you don't know about right angles, check out Chapter 2 and then come right back here.)

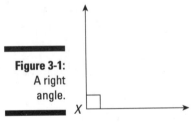

Figure 3-1:
A right
angle.

- ✔ **Major premise:** All right angles have a measure of 90 degrees.
- ✔ **Minor premise:** $\angle X$ in Figure 3-1 is a right angle.
- ✔ **Conclusion:** $\angle X$ has a measure of 90 degrees.

Here's another — this one on parallel lines (see Figure 3-2):

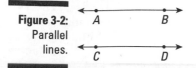

Figure 3-2:
Parallel
lines.

- ✔ **Major premise:** Lines that don't intersect are parallel.
- ✔ **Minor premise:** Line $\overleftrightarrow{AB}$ and line $\overleftrightarrow{CD}$ don't intersect.
- ✔ **Conclusion:** Lines $\overleftrightarrow{AB}$ and $\overleftrightarrow{CD}$ are parallel.

The deductive process involves segregating information into smaller and smaller groups, and the relationship between the groups must make sense in order for the conclusion to be valid.

Take a look at Figure 3-3 and then reread the preceding syllogism on parallel lines. See how the groups relate to each other? Lines is the largest group in the figure. A parallel line is a type of line, so the Parallel Lines group is inside the Lines group. Lines $\overleftrightarrow{AB}$ and $\overleftrightarrow{CD}$ don't intersect, which is the definition of parallel lines. Can't get any more inclusive in a group than that, so the Lines $\overleftrightarrow{AB}$ and $\overleftrightarrow{CD}$ group is inside the Parallel Lines group, which is inside the Lines group.

Figure 3-3:
With
deductive
reasoning,
the
relationship
between
groups must
make sense
for a
conclusion
to be valid.

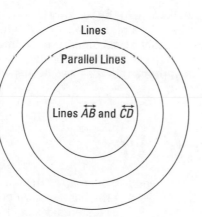

Indirect reasoning: Proving yourself wrong

The indirect method of reasoning can be a bit tricky. Unlike with the inductive and deductive methods (yup, you guessed it) you arrive at the conclusion in an indirect way. Instead of going directly to the mailbox, you first go down and around the block and through a couple of neighbors' yards.

With direct proof methods, you're looking for examples that support the premise that if p is true, then q is also true. With the indirect method of proof, you actually start by assuming that what you want to prove is false. I really don't like to encourage this type of thinking because everyone should be happy and positive, but when it comes to indirect reasoning, think negative. (Or is that contrapositive?) If you want to prove that q is false, then you actually want to prove *not q*. In geometric shorthand, *not q* looks like ~q. If you do the following four steps of indirect reasoning, you should arrive at ~q is false, meaning q is true:

1. **List the conclusion (in this case, q) that you want to prove, as well as all other possible conclusions.**

2. **Assume that the other possible conclusions are true (an easy step).**

3. **Go through the other possible conclusions one by one and show that if any were true, they'd violate a statement that's already accepted as true.**

    In other words, go through those other conclusions and prove that if any of them were true, they'd violate a given, a postulate, a theorem, or such. (Check out Chapter 1 for what those terms mean.)

4. **Draw your final conclusion.**

    Here, the only remaining conclusion is that q is the only possibility left. So it must be true.

The best application of indirect reasoning in the real world is in the courts. A defendant must be proven guilty beyond a reasonable doubt. The best way to prove that a defendant couldn't possibly have committed a crime is to present the situation that would've had to exist in order for the defendant to have committed the crime and *then* prove that it's false. That's why a good alibi is so important.

You've Gotta Justify It All

Before you can attempt a proof, you need a firm grasp of the basics to be able to progress from one step to another. Getting the basics down up front is a real time-saver in the long run. Trust me. The elementary principles covered in this section are used to justify the statements you use in proofs.

The reflexive principle: Right back atcha

You look into a mirror, and what do you see? You. Smiling, I hope. Your image is reflected back at you. Face two mirrors toward each other, and the reflected image in each mirror is identical. It's like two things looking at each other, and both see the same thing. That's what happens with the reflexive principle: In an equation, a quantity is reflected back at itself. Sometimes you might see it as identity. Reflexive or identity, either way it's the same thang.

In equation form, the reflexive principle looks like this:

$a = a$

$\angle A = \angle A$

The reflexive principle states that a quantity is equal (or congruent) to itself.

The symmetric principle: Flip-flop

How do you stop a seesaw from tipping from one side to the other? You create an equal balance. The symmetric principle creates a balance of elements on each side of an if . . . then statement. It's like bookends.

In equation form, the symmetric principle looks like this:

If $a = b$, then $b = a$

If $\angle A = \angle B$, then $\angle B = \angle A$

The symmetric principle states that an equal quantity can be reversed.

The transitive principle: The swap

You can swap things around if they're equal to each other. The transitive principle involves an "if x and y . . . then z" statement. It allows you to determine that one quantity is equal to a second quantity because of its relationship to a third quantity. All without you having to do a single bit of measuring! You can just substitute one element for another if their quantities are equal.

In equation form, it looks like this:

If $a = b$ and $b = c$, then $a = c$

If $\angle A = \angle B$ and $\angle B = \angle C$, then $\angle A = \angle C$

The transitive principle states that if two quantities are equal to the same quantity, then the two quantities are equal to each other.

The substitution principle: Just as good as the original

The substitution principle is similar to the transitive principle. But one important difference exists: The substitution principle is a lot simpler. It allows you to replace a quantity with its equivalent in any expression.

Here's an example:

$$x = 20$$
$$x + y = 30$$

You can substitute the quantity of 20 for x in the $x + y$ equation, which gives you the following:

$$20 + y = 30$$

You can then solve for y. If you don't know how, the subtraction rule can help you. (It's covered a little later in this chapter.)

The substitution principle states that a quantity may be substituted for its equivalent in an expression.

The addition rule

With the addition rule, if you add a quantity to one side, you have to add the same quantity to the other side to keep things even.

In equation form, it can take two forms that both translate to the same thing even if they look a bit different. The first form is this:

If $a = b$ and $c = d$, then $a + c = b + d$

To demonstrate the second form, I'm adding some numbers. I'm sure you'll recognize this type of equation from algebra:

$$x - 5 = 7 \text{ (solve for } x)$$
$$x - 5 + 5 = 7 + 5$$
$$x = 12$$

Equal quantities may be added to both sides of an equation. You need to get rid of the "–5" on the left side of the equation so add "+5" to the left side to zero it out. You need to keep a balance. What you do to one side, you gotta do to the other. So, if you add 5 to left side of the equation you gotta add 5 to the right side of the equation. When you isolate the variable on one side, you determine what the variable is worth. In this case, $x = 12$.

The subtraction rule

The subtraction rule follows the same course as the addition rule. Subtract from one side, subtract the same amount from the other.

In equation form, the subtraction rule looks like this:

If $a = b$ and $c = d$, then $a - c = b - d$

And this:

$$x + 5 = 7 \text{ (solve for } x)$$
$$x + 5 - 5 = 7 - 5$$
$$x = 2$$

Equal quantities may be subtracted from both sides of an equation.

The multiplication rule

In an equation, if you multiply one side of the equation by a given quantity, then you have to do the same to the other side.

In equation form, here's what you've got:

If $a = b$, then $2a = 2b$

And this:

$$x / 5 = 7 \text{ (solve for } x)$$
$$(x / 5) \times 5 = 7 \times 5$$
$$x = 35$$

When equal quantities are multiplied by an equal quantity, their products are equal.

The division rule

The division rule is similar to the multiplication rule, but instead of multiplying, you divide.

Here's one equation form:

If $a = b$ and $c = d$, then $a / c = b / d$

And this:

$$x \times 5 = 7 \text{ (solve for } x\text{)}$$
$$(x \times 5) / 5 = 7 / 5$$
$$x = 7 / 5 \text{ (or } 1\tfrac{2}{5} \text{ if you can't leave it as an improper fraction)}$$

And this:

$$6x = 12 \text{ (solve for } x\text{)}$$
$$6x / 6 = 12 / 6$$
$$x = 2$$

When equal quantities are divided by an equal quantity (other than zero), their quotients are equal.

The roots and powers rules

The roots and powers rules are pretty straightforward.

Here's the roots rule in equation form:

If $a = b$, then $\sqrt{a} = \sqrt{b}$

Positive square roots of equal quantities are equal.

And here's the powers rule in equation form:

If $a = 7$, then $(a)^2 = (7)^2$

That is, if $a = 7$, then $(a)^2 = 49$

The squares of equal quantities are equal.

"Statements rule!"

Actually, it's statement rules. I know — what a letdown. The following statement rules are a guide to let you know what is and isn't possible when you're doing geometry:

- ✔ A theorem must hold up under all tests.
- ✔ A statement is either true *or* false. Nothing can be true and false at the same time. Either it is or it isn't — but not both.
- ✔ Just because "if a then b" is true, that doesn't mean "if b then a" is true.
- ✔ If p then q is true. It's not always the case that "if $\sim p$ then $\sim q$" is also true.

Chapter 4

The Formal Proof Dressed Down

I need to solve a crime mystery. I survey the crime scene, gather the facts, and write them down in my memo pad. To solve the crime, I take the known facts and, step by step, show who committed the crime. I conscientiously provide supporting evidence for each statement I make.

Amazingly, this is the same process you use to solve a proof. This chapter walks you through the whole shebang.

The Formal Steps for Proofs: It's Just Like Doing the Waltz

Actually, this waltz is a little different from your traditional waltz. This one has *five* steps. But I'm sure that with a bit of practice, you'll get the steps down flawlessly.

1. **Get or create the statement of the theorem.**

   The statement is what needs to be proved in the proof itself. Sometimes this statement may not be on the page. That's normal, so don't fret if it's not included. (And don't fret if you see proofs in this book without the statement line included.) If it's missing in action, you can create it by changing the geometric shorthand of the information provided into a statement that represents the situation. Check out the section "Dissecting If . . . Then Statements," later in this chapter, for the details.

2. State the given.

The *given* is the hypothesis and contains all the facts that are provided. The given is the *what*. What info have you been provided with to solve this proof? The given is generally written in geometric shorthand in an area above the proof.

I know this is where traditional waltz steps end, but this waltz has a couple more. Be careful not to step on your partner's toes.

3. Get or create a drawing that represents the given.

They say a picture is worth a thousand words. You don't exactly need a thousand words, but you do need a good picture. When you come across a geometric proof, if the artwork isn't provided, you're going to have to provide your own. Look at all the information that's provided and draw a figure. Make it large enough that it's easy on the eyes and that it allows you to put in all the detailed information. Be sure to label all the points with the appropriate letters. If lines are parallel, or if angles are congruent, include those markings, too.

4. State what you're going to prove.

The last line in the statements column of each proof matches the *prove* statement. The prove is where you state what you're trying to demonstrate as being true. Like the given, the prove statement is also written in geometric shorthand in an area above the proof. It references parts in your figure, so be sure to include the info from the prove statement in your figure.

5. Provide the proof itself.

The proof is a series of logically deduced statements — a step-by-step list that takes you from the given; through definitions, postulates, and previously proven theorems; to the prove statement.

✔ The given is not necessarily the first information you put into a proof. The given info goes wherever it makes the most sense. That is, it may also make sense to put it into the proof in an order other than the first successive steps of the proof.

✔ The proof itself looks like a big letter T. Think T for *theorem* because that's what you're about to prove. The T makes two columns. You put a *Statements* label over the left column and a *Reasons* label over the right column.

✔ Think of proofs like a game. The object of the proof game is to have all the statements in your chain linked so that one fact leads to another until you reach the prove statement. However, before you start playing the proof game, you should survey the playing field (your figure), look over the given and the prove parts, and develop a plan on how to win the game. Once you lay down your strategy, you can proceed statement by statement, carefully documenting your every move in successively numbered steps. Statements made on the left are numbered and correspond to similarly numbered reasons on the right. All statements you make must refer back to your figure and finally end with the prove statement. The last line under the Statements column should be exactly what you wanted to prove.

✔ The principles of equality are frequently used as reasons to justify statements made in your proof. Examples are the reflexive, symmetric, and transitive principles, as well as the substitution, addition, subtraction, and multiplication rules. (Check out Chapter 3 for the nitty-gritty details on these elementary principles.) I know that when I solve proofs, the transitive and substitution principles are particularly handy because they allow me to swap information within the proof.

Take a look at Proof 4-1, which is really just an empty sample proof. It shows the basic setup as outlined in the five steps for proofs, as well as the general layout of a proof itself. Don't worry about it being empty. I will be filling it in, in short order.

Statement:

Given:

Prove:

	Statements		Reasons
	1)		1)
	2)		2)
Proof 4-1:	3)		3)
A proof	4)		4)
layout.	5)		5)

Playing the Proofing Game

The best way to get the hang of proofs is to study a bunch of them. One after another. In the case of proofs, quantity eventually leads to quality — and proficiency. In this section, I show you proofs for lines and angles. More complex proofs are provided in other chapters.

For Proof 4-2 through Proof 4-4, I'll provide a little running commentary so you'll get the hang of it faster.

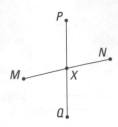

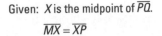

Given: X is the midpoint of $\overline{PQ}$.

$$\overline{MX} = \overline{XP}$$

Prove: $\overline{QX} = \overline{MX}$

Statements	Reasons
1) X is the midpoint of $\overline{PQ}$.	1) Given.
2) $\overline{QX} = \overline{XP}$	2) The definition of a midpoint. A midpoint is halfway between the two endpoints of a segment (see Chapter 2).
3) $\overline{MX} = \overline{XP}$	3) Given.
4) $\overline{QX} = \overline{MX}$	4) Transitive. If two quantities are equal to the same quantity, then they are equal to each other (See Chapter 3).

Proof 4-2: The Midpoint Proof

The first thing I do when I come across a proof is take a quick look at the figure (if there is one). If there isn't a figure, I draw one. It definitely makes everything more concrete. Next, I look at the Given. The information given in Proof 4-2 tells me that X is the midpoint of $\overline{PQ}$. I also know from the given that $\overline{MX}$ is equal in length to $\overline{XP}$. I then look back at the figure while keeping this information in mind. I associate the given information to the figure. Next, I look at the Prove statement and determine how it relates to the figure. I need to show that $\overline{QX} = \overline{MX}$. Now, comes the challenging part: developing a game plan of how to get from the Given to the Prove. If you are having trouble making the connection go back to Chapters 1, 2, and 3. There you'll find the basics, a virtual do-it-yourself manual on proofs.

If I can show that $\overline{QX} = \overline{XP}$ then I can use the transitive principle (from Chapter 3) to swap equal parts and arrive at $\overline{QX} = \overline{MX}$. Using the definition of midpoint and the information in the Given, I can establish that $\overline{QX} = \overline{XP}$. Then it's just a short skip and a jump to $\overline{QX} = \overline{MX}$ using the Given information of $\overline{MX} = \overline{XP}$ and the transitive principle. Using this information, I am able to cancel out the common segment ($\overline{XP}$) that both $\overline{QX}$ and $\overline{MX}$ are equal to and I get $\overline{QX} = \overline{MX}$.

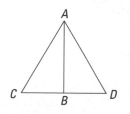

Given: $\overline{AB}$ is a segment bisector of $\overline{CD}$.

Prove: $\overline{CD} = 2BD$

	Statements		Reasons
	1) $\overline{AB}$ is a segment bisector of $\overline{CD}$.	1)	Given.
	2) $\overline{CB} = \overline{BD}$	2)	The definition of a segment bisector. A segment bisector divides a segment into two equal segments (see Chapter 2).
	3) $\overline{CD} = 2\overline{BD}$	3)	Substitution. If two quantities are equal, then they can be substituted for each other (see Chapter 3).

Proof 4-3:
The Equal
Segments
Proof

Look at the figure in Proof 4-3, then look at the Given, and then back at the figure. Keeping this information in mind, review the Prove statement. Determine how to get from the Given to the Prove. I know I need to show that $\overline{CB} = \overline{BD}$ in order to be able to get to the Prove. I start with my Given statement "$\overline{AB}$ is a segment bisector of $\overline{CD}$." Then I translate it into something I can use: $\overline{CB} = \overline{BD}$, just what the doctor ordered. I now know that $\overline{BD}$ is half the length of $\overline{CD}$, so twice (2 times) the length of $\overline{BD}$ equals the length of $\overline{CD}$. That explains how I get $\overline{CD} = 2\overline{BD}$.

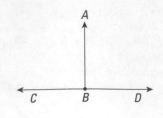

Given: $\overleftrightarrow{CD} \perp \overrightarrow{AB}$

$\overleftrightarrow{CD}$ and $\overrightarrow{AB}$ intersect at B.

Prove: $\angle ABC$ is a right angle.

Statements	Reasons
1) $\overleftrightarrow{CD} \perp \overrightarrow{AB}$.	1) Given.
2) $m\angle ABC = 90°$	2) The definition of perpendicular lines. Perpendicular lines form a 90° angle at the point of intersection (see Chapter 2).
3) Right angles have a measure of 90°.	3) The definition of a right angle (see Chapter 2).
4) $\angle ABC$ is a right angle.	4) Substitution (see Chapter 3).

Proof 4-4:
The Perpendicular Lines Proof

This is the last one I am going to do a commentary on before I turn you loose. Look at the figure in Proof 4-4, then look at the Given, and then back at the figure. Keeping this information in mind, review the Prove statement. Determine how to get from the Given to the Prove. (These steps don't ever change.) I see by the Given, that I must use information about perpendiculars to arrive at declaring $\angle ABC$ is a right angle. This sets into motion a series of logical steps. From the Given, I know that $\overleftrightarrow{CD}$ is perpendicular to $\overrightarrow{AB}$. I know that perpendiculars form 90-degree angles. I know that all right angles have a measure of 90 degrees. Because $\angle ABC$ is formed by a perpendicular, it is has a measure of 90 degrees and is therefore a right angle. See how you just get on a roll, one small logical step at a time and before you know it, you're done!

Now walk through Proof 4-5 through Proof 4-8 on your own. You're getting the hang of it!

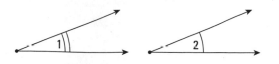

Given: m∠1 = 35°

m∠2 = 35°

Prove: m∠1 = m∠2

	Statements	Reasons
Proof 4-5: The Equal Angles Proof	1) m∠1 = 35°	1) Given.
	2) m∠2 = 35°	2) Given.
	3) m∠1 = m∠2	3) Transitive.

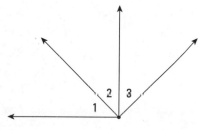

Given: m∠2 + m∠3 = 90°

m∠1 = m∠3

Prove: m∠1 + m∠2 = 90°

	Statements	Reasons
Proof 4-6: The Equal Angles Proof	1) m∠2 + m∠3 = 90°	1) Given.
	2) m∠1 = m∠3	2) Given.
	3) m∠1 + m∠2 = 90°	3) Substitution (see Chapter 3).

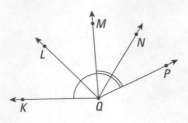

Given: $\overrightarrow{QL}$ bisects $\angle KQM$.

$\overrightarrow{QN}$ bisects $\angle MQP$.

Prove: $m\angle KQL + m\angle NQP = m\angle LQN$

Statements	Reasons
1) $\overrightarrow{QL}$ bisects $\angle KQM$.	1) Given.
2) $m\angle KQL = m\angle LQM$	2) The definition of an angle bisector. $\overrightarrow{QL}$ bisects $\angle KQM$ into two equal angles (see Chapter 2).
3) $\overrightarrow{QN}$ bisects $\angle MQP$.	3) Given.
4) $m\angle MQN = m\angle NQP$	4) Same as #2.
5) $m\angle KQL + m\angle NQP = m\angle LQM + m\angle MQN$	5) Addition. Equal quantities added together are equal to each other.
6) $m\angle LQM + m\angle NQM = m\angle LQN$	6) Angle Addition. A whole is the sum of its parts (see Chapter 2).
7) $m\angle KQL + m\angle NQP = m\angle LQN$	7) Substitution. Equal quantities can replace each other (see Chapter 3).

Proof 4-7:
The Angle
Bisector
Proof

Given: ∠1 is a supplement of ∠2.

∠3 is a supplement of ∠4.

m∠2 = m∠4

Prove: m∠1 = m∠3

Statements	Reasons
1) ∠1 is a supplement of ∠2.	1) Given.
2) ∠3 is a supplement of ∠4.	2) Given
3) m∠1 + m∠2 = 180°	3) The definition of supplementary angles (see Chapter 2).
4) m∠3 + m∠4 = 180°	4) Same as #3.
5) m∠2 = m∠4	5) Given.
6) m∠1 + m∠2 = m∠3 + m∠4	6) Addition. Equal quantities added together are equal to each other (see Chapter 3).
7) m∠1 + m∠2 = m∠3 + m∠2	7) Substitution (see Chapter 3).
8) m∠1 = m∠3	8) Subtraction. Equal quantities can be subtracted from each side of an equation.

Proof 4-8:
The
Supplements
of Angles
Proof

Proofing Indirectly

Sometimes you just can't get from the Given part to the Prove part directly (using the method of deductive reasoning — see Chapter 3). In these situations, you can try the indirect method. With the indirect method, you show that the hypothesis is true by assuming that the conclusion you're trying to prove is not true. Then you go about proving that the hypothesis must be the only choice that could possibly be true.

Yes, I should probably add a little to that description so it makes more sense. Let me try again. Here are the steps for solving a proof indirectly:

1. **Assume that the conclusion, or the Prove statement, is the opposite of what's stated.**

   In other words, assume that it's false.

2. **Show that according to the information in the Given statement, the conclusion could not be false.**

   It may take a series of statements to reach this point.

3. **Once you've determined that the conclusion couldn't be false, conclude that it's true because it being false contradicts a known fact.**

Proofs 4-9 and 4-10 are proved indirectly. Look at the proofs, read the information about them, and then look at the proofs again. See how the statements follow a valid progression of reasoning, even though you're trying to prove that something is not true?

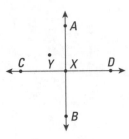

Given: $\overleftrightarrow{AB}$ and $\overleftrightarrow{CD}$ are intersecting lines.

Prove: $\overleftrightarrow{AB}$ and $\overleftrightarrow{CD}$ intersect at point X.

Indirect proof list of possibilities:

   1) $\overleftrightarrow{AB}$ and $\overleftrightarrow{CD}$ intersect at two or more points. (Assume first that this statement is true and add another point to the figure, say Point Y, at a location different from Point X.)

   2) $\overleftrightarrow{AB}$ and $\overleftrightarrow{CD}$ intersect at only point X.

Statements	Reasons
1) $\overleftrightarrow{AB}$ and $\overleftrightarrow{CD}$ are intersecting lines.	1) Given.
2) $\overleftrightarrow{AB}$ and $\overleftrightarrow{CD}$ intersect at point X and point Y.	2) This assumption violates the postulate stating that one and only one line can be drawn between two points (see Postulate 1-1).
3) $\overleftrightarrow{AB}$ and $\overleftrightarrow{CD}$ intersect at only point X.	3) Because Possibility 1 is false, you assume that Possibility 2 is true.

Proof 4-9:
The
Intersecting
Lines Proof

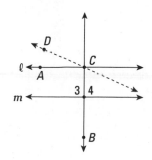

Given: $\ell \parallel m$

Prove: $m\angle ACB = m\angle 4$

Indirect proof list of possibilities:

1) The measure of $\angle ACB$ is not equal to the measure of $m\angle 4$. (Assume first that this statement is true.)

2) The measure of $\angle ACB$ is equal to the measure of $\angle 4$.

Statements	Reasons
1) $\ell \parallel m$	1) Given.
2) $m\angle ACB \neq m\angle 4$	2) If this assumption is true, then another line can be drawn through point C such that alternate interior angles are congruent.
3) $m\angle ACD = m\angle 4$	3) If Possibility 1 (from my possibility list for this proof) is true, then it violates the Given because one and only one line can be drawn through point C and be parallel to line m. The assumption that $m\angle ACB = m\angle 4$ must therefore be true (see Chapter 2).

Proof 4-10:
The Parallel
Lines Proof

I'll run through Proof 4-9 because indirect reasoning can be a bit awkward. First, review the Given information and the figure just like with direct reasoning proofs earlier in this chapter. The trick with indirect reasoning is to show that the Prove is true by showing that it is the only possible solution. This means you get to make stuff up and stick it on your figure in an attempt to show the Prove is false. For Proof 4-9, the original Prove states that $\overleftrightarrow{AB}$ and $\overleftrightarrow{CD}$ intersect at point X. If I assume the opposite is true, then I am going to try to prove that $\overleftrightarrow{AB}$ and $\overleftrightarrow{CD}$ intersect at two or more points. I add point Y to the figure in an attempt to show that two lines can intersect at point X and point Y. This is crazy talk, though, because I know that lines intersect at only one point because that's what Postulate 1-1 says. I have just shown that $\overleftrightarrow{AB}$ and $\overleftrightarrow{CD}$ must intersect only at point X because two lines intersecting at two or more points violates a postulate and this is not acceptable. I know it may seem counter-intuitive, but this is the way it's done.

Try Proof 4-10 yourself. Start by trying to prove that the opposite of the Prove statement is true while still using the information in the Given. When you prove that the opposite of the Prove statement or any other possible alternatives couldn't be true, you must then conclude that the original Prove is true if it is the only viable conclusion left that you have not been able to prove false.

A known fact to be used as a reason can come from the Given, a definition, a postulate, or a previously proven theorem.

Dissecting If . . . Then Statements

In proofs, sometimes the statement of the theorem is printed on the page. Sometimes it's not. So when you come across a proof that's missing the statement, you can create it from the Given and the Prove parts of the proof. Because Given and Prove statements are in geometric shorthand and usually contain symbols, you have to translate them into a sentence — an *if . . . then* sentence. Minding your p's and q's, the information looks something like this: *if p, then q.* This kind of statement is known as a *conditional statement.* If the *if* condition is true, then the *then* clause is true. The *if* clause of the sentence represents the Given, and the *then* clause represents the Prove statement. When conditional statements are written in two clauses — an independent (then) clause and a dependent (if) clause — identifying the hypothesis and conclusion is easy.

However, conditional statements can take other forms. Sometimes the *then* clause may be omitted, or the *if* clause may appear at the end of the sentence. Other times the statement may be in the form of a simple sentence that doesn't contain the words *if* or *then.* In this case, the subject of the sentence is the hypothesis, and the predicate is the conclusion. Regardless of how the information is written, a conditional statement is classified as being either true or false.

Look at Proofs 4-11 through 4-13 to get to know how to translate *if . . . then* statements and how to complete the proofs that go along with them.

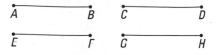

Statement: If $\overline{AB}$ and $\overline{CD}$ are equal, $\overline{EF}$ and $\overline{GH}$ are equal, $\overline{CD}$ and $\overline{GH}$ are equal, then $\overline{AB}$ and $\overline{EF}$ are equal.

Given: $\overline{AB} = \overline{CD}$
$\overline{EF} = \overline{GH}$
$\overline{CD} = \overline{GH}$

Prove: $\overline{AB} = \overline{EF}$

Proof 4-11:
The Equal Angles Proof

Statements	Reasons
1) $\overline{AB} = \overline{CD}$	1) Given.
2) $\overline{EF} = \overline{GH}$	2) Given.
3) $\overline{CD} = \overline{GH}$	3) Given.
4) $\overline{EF} = \overline{CD}$	4) Transitive (see Chapter 3).
5) $\overline{AB} = \overline{EF}$	5) Transitive (see Chapter 3).

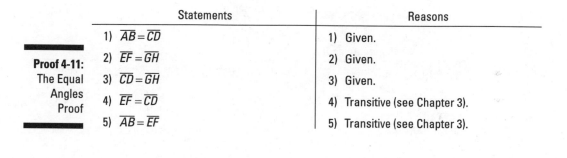

Statement: If $\angle ABC$ and $\angle EFG$ are right angles, then $m\angle ABC = m\angle EFG$.

Given: $\angle ABC$ and $\angle EFG$ are right angles.

Prove: $m\angle ABC = m\angle EFG$

Proof 4-12:
The Right Angle Proof

Statements	Reasons
1) $\angle ABC$ and $\angle EFG$ are right angles.	1) Given.
2) $m\angle ABC = 90°$ and $m\angle EFG = 90°$	2) Right angles have a measure of 90°.
3) $m\angle ABC = m\angle EFG$	3) Equal quantities are equal.

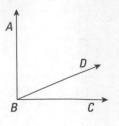

Statement: If $\overrightarrow{BA}$ is perpendicular to $\overrightarrow{BC}$, then $\angle ABD$ is complementary to $\angle DBC$.

Given: $\overrightarrow{BA} \perp \overrightarrow{BC}$

Prove: m$\angle ABD$ is complementary to m$\angle DBC$.

Proof 4-13:
The Complementary Angles Proof

Statements	Reasons
1) $\overrightarrow{BA} \perp \overrightarrow{BC}$	1) Given.
2) $\overrightarrow{BA}$ and $\overrightarrow{BC}$ form a right angle.	2) Perpendicular lines form right angles (see Chapter 2).
3) m$\angle ABC = 90°$	3) Right angles have a measure of 90° (see Chapter 2).
4) m$\angle ABC = $ m$\angle ABD + $ m$\angle DBC$	4) Angle Addition (see Chapter 2).
5) m$\angle ABD + $ m$\angle DBC = 90°$	5) Substitution (see Chapter 3).
6) m$\angle ABD$ is complementary to m$\angle DBC$.	6) Complimentary angles equal 90° (see Chapter 6).

Part III
It Takes All Shapes and Sizes

"For the next month, instead of practicing on a baseball diamond, we'll be practicing on a baseball trapezoid. At least until everyone passes the geometry section of the GRE test."

In this part . . .

If you're like me, you're always ready for dessert. I know I could use a large slice of chocolate cake *right* now. Sorry, but it's time for the main course instead. This part contains the meat of geometry. And if you're a vegetarian, no worries. This part contains the potatoes, too.

Triangles, quads, circles, and polygon stuff out the whazoo — that's what's on the menu. Dinner is served.

Chapter 5

Polygons as Appetizers (A Taste of Closed Figures)

I know, I know. You were hoping for a cooking lesson given the title of this chapter. Sorry — no delectable treats here. Just a foray into the exciting world of geometrical shapes. In this chapter, I introduce you to the basics of closed figures — polygons in general. Polygons give you a good taste of what geometry is all about: Lines and angles are mixed together to form closed areas of all different shapes and sizes. (**Note:** Triangles and quadrilaterals, the most common polygons, have their own chapters because I have a lot to say about them.)

"So a Polygon Is What, Exactly?"

Well, for one thing, it's the bottom line when your pet parrot flies the coop. For another, it's a geometrical shape — one that has at least three sides. Each side of a polygon is a line segment that touches a neighboring line segment at its endpoint. Segments that touch at the endpoint are called *consecutive sides*. Each segment touches another segment only once. There's no overlap and no opening. And because the figure is closed, where you start drawing the first line of your figure is where you stop drawing the last line of your figure.

Think of a closed figure as a pigpen. If any break occurs in the fence, all the pigs will get loose. So the goal when drawing a polygon is to make sure that the pigs don't get loose (see Figure 5-1).

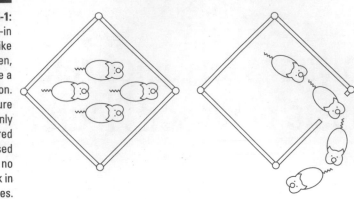

Figure 5-1:
A fenced-in
area, like
this pigpen,
can be a
polygon.
A figure
is only
considered
to be closed
if there's no
break in
the sides.

The name game

You can refer to many polygons by specific names. For example, a polygon must have at least three sides, so you can probably guess what a three-sided polygon is called. Yep, a triangle. The triangle has the least number of sides that a polygon can have and still be considered a polygon.

Polygon names are a direct reflection of the number of sides a polygon has. So a couple of other polygon names are the quadrilateral and the pentagon (see Figure 5-2).

If you're one of those folks who get a big charge out of at-a-glance setups and stats, then Table 5-1 is for you. It lists the names of the most commonly referred to polygons, as well as the number of sides each has. Enjoy.

Figure 5-2:
A quadri-
lateral is a
four-sided
polygon,
and a
pentagon is
a five-sider.

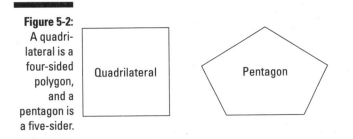

Table 5-1:	The Names of Common Polygons
Name of Polygon	*Number of Sides*
Triangle	3
Quadrilateral	4
Pentagon	5
Hexagon	6
Heptagon	7
Octagon	8
Nonagon	9
Decagon	10
Dodecagon	12

If you can't find the name of a 37-sided polygon anywhere, it's because . . . . Well, first, it's because it's not that common. And, second, it's because after 12 sides, the polygon naming thing gets sketchy. After 12 sides, polygons are generically referred to as *n*-gons. So your 37-sided polygon is a 37-gon. Awkward, yes. But it's simple and it works.

How angles come into play

Not only do polygons have sides, but they also obviously have angles that are formed where the sides meet. The funny thing is that even though polygons are commonly named by the number of their sides, the word *polygon* is a combination of two Greek words that give reference to it having many angles — *poly* meaning *many* and *gon* meaning *angle*. So *polygon* actually means "many angled." Strictly speaking, though, a polygon has both — many sides and many angles.

For simplicity's sake, you can't argue that counting sides is easier than sitting there counting angles. But with polygons, it doesn't really matter whether you count sides or angles because any given polygon has the same number of both. Take an octagon, for example (see Figure 5-3). Just as an octopus has eight tentacles, an octagon has eight sides *and* eight angles.

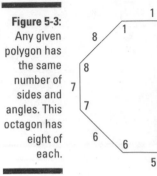

Figure 5-3:
Any given
polygon has
the same
number of
sides and
angles. This
octagon has
eight of
each.

So, in general, a polygon is named by the number of its sides. But what happens if you have two polygons that are both hexagons, and you want to distinguish between the two of them? In this case, you can name the polygons by their angles. An angle is formed at a point where the sides of a given polygon collide, and each point (where an angle is formed) is called a *vertex*. Each vertex is labeled with a unique letter to distinguish it from the other *vertices* (plural for vertex) in the polygon. Once all the vertices are named, you can refer to the polygon by using all its letters. Pick one letter to start with and write down the letter given to *each* vertex until you get all the way around the polygon. You have to choose whether you want to move clockwise or counter-clockwise around the figure. It doesn't matter which way you go; the only rule here is that you can't skip around. If you start naming in a clockwise direction, don't do the hokeypokey and turn yourself around and change to counter-clockwise midstream. If you go with clockwise, *stay* with clockwise until you use the letter for each vertex you encounter while making your way around the polygon. In Figure 5-4, for example, the polygons are referred to as polygon *ABCDEF* and polygon *RSTUVW*.

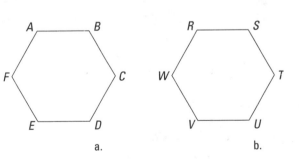

Figure 5-4:
Naming
polygons by
their angles:
polygon
ABCDEF
(Figure a)
and polygon
RSTUVW
(Figure b).

A regular polygon — that's with milk and sugar, right?

Not only can polygons be classified by the number of sides they have and by their angles, but they can also be grouped according to some of their qualities. Polygons can have three personality characteristics: equilateral, equiangular, regular.

In an *equilateral* polygon, all sides are equal and there's at least one nonsimilar angle. In an *equiangular* polygon, all angles are equal and at least one side doesn't match the length of the others. A *regular* polygon is both equilateral and equiangular; it has total symmetry — equal sides and equal angles.

Looking Inside a Polygon

Polygons are big cons. They fool you into thinking that they are nothing more than a plain, old shape with not much going on. It's time to expose polygons for what they really are by tearing down their façade and looking at them from the inside.

Is it convex or concave?

The first thing you should do is determine what type of con a particular polygon is; con-vex or con-cave. It's always nice to know what you're up against. Most polygons you'll encounter in geometry will be of the convex variety. But it's not unheard of, on some rare occasions, to come across the notorious species known as the concave polygon. How do you tell the difference? The first step is to understand what sets these two species of polygons apart from each other. Usually, you can tell by just looking at a polygon.

If you follow the sides clockwise around a *convex* polygon, you always turn right at each vertex (see Figure 5-5).

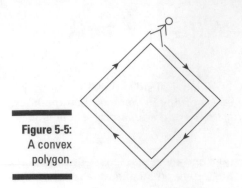

Figure 5-5:
A convex
polygon.

With a *concave* polygon, you turn left at a vertex in at least one instance. The second you encounter a vertex where you turn left instead of right, you know that the polygon is concave (see Figure 5-6). A concave polygon may look as if it has a dent in it or is caved in. The reason is that one (or more) of its angles (located at a vertex) has a measure of more than 180°. So the major difference between convex and concave polygons is that with convex polygons, each of the angles (located at a vertex) has a measure of between 0° and 180°.

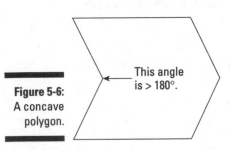

This angle
is > 180°.

Figure 5-6:
A concave
polygon.

On your geometrical expeditions, you'll encounter mostly convex polygons, so the information given in this book will generally pertain to them as opposed to concave polygons.

It's all in the diagonal

The sides of a polygon are consecutive line segments that form consecutive vertices in a closed figure. *Diagonals* are lines that cross through the interior of a polygon and connect nonconsecutive vertices of the polygon. Blech. Just look at Figure 5-7 to see what I mean.

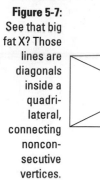

Figure 5-7:
See that big fat X? Those lines are diagonals inside a quadri-lateral, connecting noncon-secutive vertices.

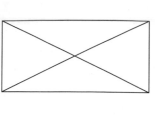

The number of diagonals that can be drawn within any given polygon is directly related to the number of sides that the polygon has. Take a pentagon. Pick one vertex (for example, vertex A) and draw one line from vertex A to each of the other vertices in the polygon. Some of the lines are diagonals, and some are not. The lines that are not sides of the figure are diagonals (see Figure 5-8). So you can conclude that a quadrilateral, for example, has two diagonals. But what about the pentagon? It has two diagonals just for vertex A. You can simply multiply the number of sides ($n = 5$) by the number of diag-onals you can draw for a single vertex — in this case, 2. A pentagon has ten lines that can be drawn between vertices. Half of these lines are duplicates, which leaves you with five unique lines, or diagonals, connecting the noncon-secutive vertices. In other words,

5 sides (2 diagonals for one side = ½ (10 lines for 5 sides), or 5 diagonals

Figure 5-8:
If you draw lines from vertex A to all the other vertices, the diagonals are the lines that aren't sides of the pentagon ($\overline{AC}$ and $\overline{AD}$).

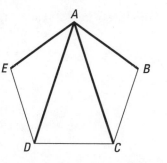

You can simplify things dramatically by using a real formula to determine the number of diagonals. That way, you don't have to draw a figure for every polygon to determine its number of diagonals. The formula for determining the total number of unique diagonals (D) in a polygon is

$$D = ½ n(n - 3)$$

Theorem 5-1: The total number of unique diagonals (D) for a polygon of n sides is $D = ½ n(n - 3)$.

Translation: Use the information you have about the number of sides of your polygon and plug it into the formula. For n, substitute the number of sides in your polygon and then solve for D. D is the total number of diagonals.

So if you use the diagonals formula with the pentagon example, you get this pretty scene:

$$D = ½ 5(5 - 3)$$
$$D = ½ 5(2)$$
$$D = ½ 10$$
$$D = 5$$

Figuring the sum measures of interior angles

I hear ya: "Doing formula stuff is fine and dandy, but why do I *need* diagonals?" Have you noticed that diagonals divide the interior area of a polygon into a number of triangles? Because a straight line is 180°, it follows that the sum of a triangle's three *interior angles* (angles that are inside a figure) also equals 180° (see Figure 5-9). I go into the specifics of triangles in Chapter 6. The important point here is that I'm introducing you to Theorem 5-2:

Theorem 5-2: The sum of the measures of the interior angles of a triangle is 180°.

Translation: If you add the measures of the three angles of a triangle, they equal 180°.

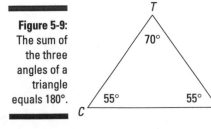

Figure 5-9:
The sum of the three angles of a triangle equals 180°.

In Figure 5-9, the sum measure of the three angles of △TNC equals 180° because 70° + 55° + 55° = 180°.

You use diagonals to determine the sum measure of the interior angles of a polygon. Any convex polygon with n sides can be divided into $n - 2$ triangles. A rectangle can be divided into two triangles. If the sum measure of the interior angles is 180° for each triangle, then the measure of a rectangle's interior angles equals 180(2), or 360°. In Figure 5-10, for example, the sum of the measures of the four angles of a quadrilateral equals 360°.

Figure 5-10: The sum of the measures of the interior angles of a quadrilateral is 360°.

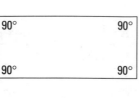

Theorem 5-3: The sum of the measures of the interior angles of a quadrilateral is 360°.

Translation: If you add the measures of the four angles of a quadrilateral, they equal 360°.

You can also determine the sum of the interior angles by using a formula. Yes, another formula. (There are two certainties in geometry: definitions and formulae. Ain't life grand?) To find the sum of the interior angles of any polygon, you need only know the number of sides of that polygon. You can determine the sum of the interior angles by using the following formula:

$$S = (n - 2)\ 180°$$

Note that $n - 2$ is the number of unique triangles formed by the diagonals with no overlapping areas. Multiply this number of unique triangles by the sum of the interior angles of a triangle (180°), and presto, you have the total number of degrees of the interior angles of that polygon.

Theorem 5-4: If a polygon has n sides, the formula for the sum of its interior angles is $S = 180°(n - 2)$.

Translation: Use the information you have about the number of sides of your polygon and plug it into the formula. For n, substitute the number of sides in your polygon and then solve for S. S is the sum of the interior angles of a convex polygon with n sides.

Want a quickie example? If you have a heptagon, you can find the sum of its interior angles by substituting a 7 for the letter n in the equation and then solving for S:

$S = 180(7 - 2)$

$S = 180(5)$

$S = 900°$

The sum of the interior angles of a heptagon is equal to 900°. As n (the number of sides) changes, so does the sum of the interior angles (see Table 5-2).

Table 5-2:	Sums of Interior Angles of Various Polygons
Number of Sides	*Sum of Interior Angles*
3	180°
4	360°
5	540°
6	720°
7	900°

Even when you know the sum of the interior angles of a polygon, that sum doesn't indicate the measure of any single angle.

A regular polygon is the exception to the rule that the sum of the interior angles doesn't indicate the measure of any single angle. For example, if a given heptagon is a regular polygon, then by definition, all its interior angles are equal. To find the measure of each interior angle, you divide the sum of the interior angles by the number of sides in the polygon. A heptagon has seven sides, and the sum of its interior angles is 900°, so the measure of any one angle in this polygon is 900°/7, or 128.57° per angle. Now you can die happy, right? You're welcome.

Theorem 5-5: In a regular polygon, the measure of an interior angle is equal to $180°(n - 2)/n$.

Translation: If a polygon is regular, the measure of each interior angle is equal to $180°(n - 2)/n$.

Figuring the sum measures of exterior angles
Although the sum of the measure of interior angles may depend on the number of sides of the polygon, the sum of the exterior angles doesn't. Regardless of the number of sides, if you use an exterior angle from each

vertex, the sum of the exterior angles is always 360°. For example, the polygon in Figure 5-11 is a regular polygon. It has five equal sides. Each of the exterior angles has a measure of 72°. Seventy-two multiplied by 5 equals 360°.

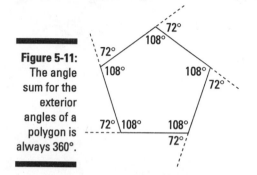

Figure 5-11:
The angle sum for the exterior angles of a polygon is always 360°.

Theorem 5-6: The sum of exterior angles is 360°.

Translation: If you use an exterior angle at each vertex, the sum of the exterior angles is 360°.

Theorem 5-7: In a regular polygon, the measure of each exterior angle is 360°/*n*.

Translation: If you have a regular polygon, the measure of each exterior angle is equal to 360°/*n*, where *n* is the number of sides in the polygon.

An exterior angle is formed when one of the sides of a polygon is extended through its vertex and reaches out into the exterior. An exterior angle isn't located within the confines of the polygon itself, but it does have a relationship with its adjacent interior angle. The measure of an exterior angle is supplementary to the measure of its neighboring angle that is trapped inside the polygon. In Figure 5-12, for example, angle 1 and angle 2 share a side and lie on a straight line. The measures of the two angles equal 180°.

Figure 5-12:
An exterior angle and an adjacent interior angle at a vertex are supplementary.

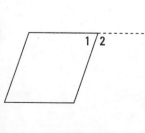

Sizing Up the Area of a Polygon

Regular polygons have lines with special meaning. These lines include the radius and the apothem.

The *radius* is a line that goes from the center of the polygon into an elbow (or vertex if you prefer the technical babble) of the polygon — splitting that angle evenly into two. When two different radii in a polygon are drawn to two consecutive vertices, a *central angle* is formed in the center of the polygon (see Figure 5-13).

Figure 5-13:
Two radii drawn to two consecutive vertices form a central angle of a regular polygon.

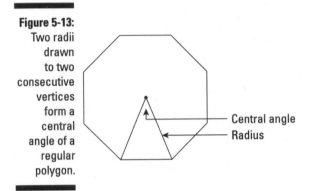

— Central angle
— Radius

Unlike the radius, which intersects an angle, an *apothem* runs from the center of the polygon straight into a flat side of the polygon. On impact, the apothem becomes a perpendicular bisector of the side it collides with (see Figure 5-14).

Figure 5-14:
An apothem of a regular polygon becomes a perpendicular bisector.

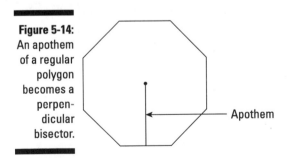

— Apothem

A slew of theorems exist for radii, central angles, and apothems of regular polygons. Here's a summary for your reading pleasure:

- ✔ *Theorem 5-8:* Radii of a regular polygon bisect the interior angle.

- ✔ *Theorem 5-9:* Central angles of a regular polygon are congruent.

- ✔ *Theorem 5-10:* Central angles of regular polygons with equal sides are congruent.

- ✔ *Theorem 5-11:* The measure of a central angle in a regular polygon is equal to 360° divided by the number of sides of the polygon.

- ✔ *Theorem 5-12:* An apothem of a regular polygon bisects the central angle (determined by the side) to which it's drawn.

- ✔ *Theorem 5-13:* An apothem of a regular polygon is a perpendicular bisector to the side it's drawn to.

You can calculate the area of a regular polygon by using the length of its apothem and the length of its perimeter: You need to survey the perimeter and determine its length. Use the information about the length of one side. Because the polygon is regular, the lengths are the same for each side. Multiply the number of sides of the polygon by the length of one side, and you get the perimeter. The area of a regular polygon is equal to one-half the product of the apothem and the perimeter.

Theorem 5-14: The formula for the area of a regular polygon is $A = \frac{1}{2}ap$, where a is the apothem and p is the perimeter.

Translation: If you have a regular polygon, plug the length of the apothem and the perimeter into the formula, and you get the area.

Take a look at Figure 5-15 for an example. The information given indicates that the length of one side of the pentagon equals 5 and that the apothem equals 6. Before you can determine the area, you must first calculate the perimeter. If the length of one side of a pentagon is 5, then the perimeter is equal to a side length of 5 multiplied by five sides. So the total perimeter of the pentagon equals 25. If you plug this information into the area formula, you get the following:

$A = \frac{1}{2}(6)(25)$

$A = \frac{1}{2}(150)$

$A = 75$

So the area of the pentagon in Figure 5-15, with the given information, is 75 square units.

Figure 5-15:
You can
determine
the area of a
pentagon by
starting with
the length of
one side
and the
length of the
apothem.

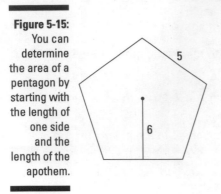

Now consider this: Just as you can add line segments and angles, you can also add areas.

Postulate 5-1: If a polygon encloses smaller, non-overlapping regions within its perimeter, then the area of that polygon is equal to the sum of the areas of the enclosed regions.

Take a look at the concave polygon in Figure 5-16. To find the total area of the figure, get the area of sections that you can easily obtain. Look closely: You can actually break the polygon into two non-overlapping rectangles. Find the area of each rectangle and then add them together. You then have the area of the whole polygon.

Figure 5-16:
In a
polygon, the
sum of the
areas of
the non-
overlapping
regions
equals the
whole area
of the
polygon.

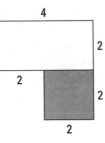

Chapter 6

Angle + Angle + Angle = Triangle

I guess it's pretty obvious how the triangle got its name. But although it may look like nothing more than a three-sided figure, I can assure you that there's more to this book than its cover. In this chapter, you discover what makes a triangle a triangle: its parts, its angles, and, of course, its character and charm.

Parts Is Parts

A triangle is a polygon. In fact, it has the least number of sides a polygon can have. Because a polygon must be a closed figure, at least three sides are needed to accomplish the closure. To be considered plane, a figure needs a minimum of three non-collinear points. Once again, the triangle just barely squeaks by.

So a triangle contains the fewest number of sides required for a polygon and the fewest number of points to be considered a plane. In this section, you take a close look at these sides and points to see what other special secrets lie behind the deceptive exterior of the triangle. The triangle's importance in geometry can no longer be cloaked.

The sides and angles

When the three sides of a triangle come together, they form the six parts of a triangle. *Six?* Yep — three sides and three interior angles that the three sides form. In Figure 6-1, the three sides are formed by line segments $\overline{AB}$, $\overline{BC}$, and $\overline{CA}$.

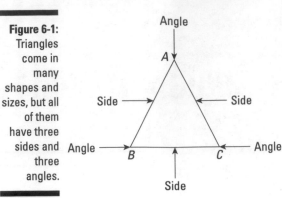

Figure 6-1:
Triangles come in many shapes and sizes, but all of them have three sides and three angles.

A *vertex* is the point at which two sides meet.

In Figure 6-1, the angles of the triangle are $\angle A$, $\angle B$, and $\angle C$, and they're located at vertices *A*, *B*, and *C*, respectively. Angles whose vertices are consecutive are called *consecutive angles,* so the three angles of the triangle in Figure 6-1 are consecutive angles.

The triangle in Figure 6-1 also has a name. And it's not Joe. A triangle is named by its vertices, so this triangle is named *ABC*. But this isn't the full name. It's missing the cute little triangle symbol — △. △*ABC* is the full name of the triangle shown in Figure 6-1.

So you can name the line segments of a triangle, the vertices, the angles, and the triangle itself. And one more thing: You can name the sides of a triangle. You do so with the lowercase version of the letter that names the opposite vertex (see Figure 6-2). Naming a side without using the name of its line segment makes identification more simple and concise. You'll be grateful for naming it this way when you start using side names in equations.

Figure 6-2:
· To identify the sides of a triangle, use the lowercase letters of the opposite vertices.

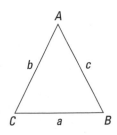

Auxiliary lines

Five lines, called *auxiliary lines,* are part of a triangle — and yet they're not. These lines include the median, midline, altitude, perpendicular bisector, and angle bisector, and they assist you when you're adding additional info to a figure or constructing a proof. You can't just add these lines to your figure willy-nilly. As with everything else, you gotta follow the rules. In particular, whenever you draw a line, it has to be determined. And I don't mean that *you* have to be determined to draw the line. If a line is *determined,* it's the only line that satisfies the set of given conditions. The information is just right. On the other hand, *underdetermined* means the information given is *under* the amount you need to narrow things down to draw only one line. Too few conditions are given, and many lines — not just one — can meet the conditions. With *overdetermined,* you have information overload. Too many given conditions exist, and they can never be satisfied by just one line. It's a situation in which doing just one thing won't please the conditions.

The median

The median doesn't provide any inherent information about the vertex it originates from. But you can use it as an indication of the halfway point (the midpoint) of the line it's being drawn to because the segments on either side of the midpoint are congruent (see Figure 6-3).

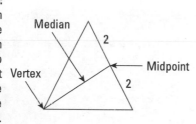

Every triangle has three medians — one from every vertex to the midpoint of the opposite side. The medians of a triangle are *concurrent,* meaning that all three lines have one point in common. The common point at which the medians meet is called the *centroid* of the triangle (see Figure 6-4). The centroid is two-thirds the distance from a vertex to the midpoint of the opposite side. So in Figure 6-4, $\overline{AD}$, $\overline{BE}$, and $\overline{CF}$ are concurrent at point X, and $\overline{AX} = 2/3\overline{AD}$, $\overline{BX} = 2/3\overline{BE}$, and $\overline{CX} = 2/3\overline{CF}$.

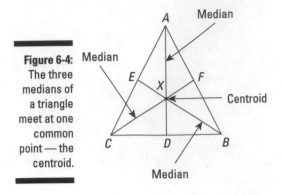

Figure 6-4: The three medians of a triangle meet at one common point — the centroid.

Theorem 6-1: The three medians of a triangle are concurrent.

Translation: There are three medians in a triangle, one for each side, and they share a common point.

Theorem 6-2: The point of concurrency of the medians of a triangle is, from any vertex, two-thirds the distance from that vertex to the midpoint of the opposite side.

Translation: All three medians meet at a point, from any vertex, that is two-thirds the distance from that vertex to the midpoint of the opposite side. (Best I could do. This one defies simplification.)

The midline

The midline connects the midpoints of two sides of a triangle (see Figure 6-5).

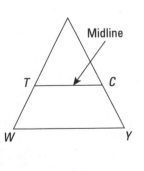

Figure 6-5: The midline is a line drawn from the midpoint of one side of a triangle to the midpoint of another side of that triangle.

Theorem 6-3: The midline of a triangle is parallel to the third side.

Translation: The midline is parallel to the side it doesn't touch.

So in Figure 6-5, $\overline{TC}$ is parallel to $\overline{WY}$.

Theorem 6-4: The midline is half as long as the third side of the triangle.

Translation: The midline is half as long as the side it doesn't touch.

So in Figure 6-5, the length of $\overline{TC}$ equals one-half the length of $\overline{WY}$.

The altitude

Think of the altitude as a measure of the triangle's height. Regardless of whether the sides of your triangle are slanted, always measure straight from the highest point. Always draw an altitude perpendicular to the opposite side (see Figure 6-6a). To do so, you sometimes may have to extend the side of your triangle (see Figure 6-6b).

Figure 6-6:
The altitude is a line drawn from the vertex perpendicular to the opposite side or side extension.

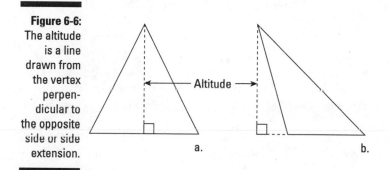

a.   b.

Theorem 6-5: The altitudes of a triangle (or the lines containing the altitudes) are concurrent.

Translation: There are three altitudes (or lines containing altitudes) for any given triangle, one for each angle, and they meet a common point.

In Figure 6-7, $\overline{AD}$, $\overline{BE}$, and $\overline{CF}$ are altitudes that are concurrent at point X.

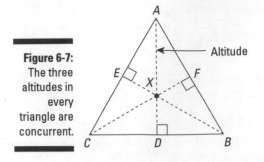

Figure 6-7:
The three altitudes in every triangle are concurrent.

Theorem 6-6: In a given triangle, the product of the length of any side and the length of the altitude drawn to that side is equal to the product of the length of any other side and the altitude drawn to that side.

Translation: If you multiply the length of any side of a triangle by the length of the altitude to that side, it's equal to the length of any other side multiplied by its altitude.

In Figure 6-7, $\overline{AD}$ and $\overline{BE}$ are altitudes, so $(\overline{CB})(\overline{AD}) = (\overline{AC})(\overline{BE})$.

The perpendicular bisector

The perpendicular bisector of a triangle is a line that forms two right angles when it meets a side. It also splits the side into two congruent segments (see Figure 6-8). If it is drawn through a side it creates four right angles, two on either side of the intersected line.

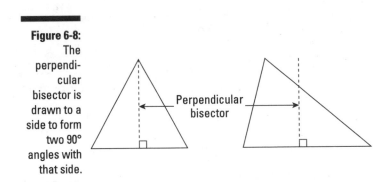

Figure 6-8:
The perpendicular bisector is drawn to a side to form two 90° angles with that side.

Perpendicular bisector

To show that a line is a perpendicular bisector, you have to prove two things: first, that the line is perpendicular to a side of the triangle and, second, that it bisects that side. To show that a line bisects another, just show that the line divides the segment into two equal segments. Next, show that the line is perpendicular to the side by using distance or degree. If you want to do it by degree, show that when a line intersects the side, it forms congruent, adjacent angles — or right angles. If you want to prove it by distance, show that the two points on the potential perpendicular are both equidistant from the end of the triangle side.

Theorem 6-7: The perpendicular bisectors of the sides of a triangle are concurrent at a point equidistant from any vertex of the triangle (see Figure 6-9).

Translation: The three perpendicular bisectors (one from each side) of a triangle meet at a single point that is the same distance from any angle of the triangle.

Figure 6-9:
The perpendicular bisectors of a triangle are concurrent.

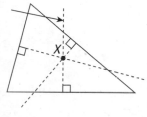

In Figure 6-9, the concurrent perpendicular bisectors intersect at point X. Point X is of equal distance from every vertex in the triangle.

The angle bisector

The angle bisector of a triangle is a line that splits an angle evenly into two congruent angles and extends to the opposite side (see Figure 6-10). The angle bisector and the median in any given triangle are generally not the same.

Figure 6-10:
An angle bisector of a triangle bisects an angle and extends to the opposite side.

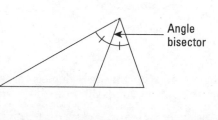

Angle bisector

Theorem 6-8: The angle bisectors of a triangle are concurrent at a point equidistant from every side of the triangle (see Figure 6-11).

Translation: The three angle bisectors (one from each angle) of a triangle meet at a single point that is the same distance from every side of the triangle.

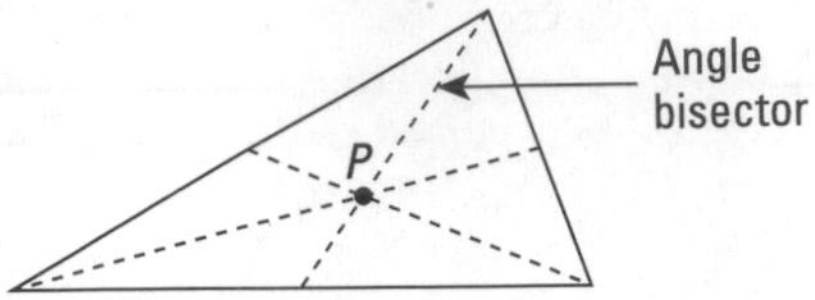

Figure 6-11:
The angle bisectors of a triangle are concurrent at a point that's of equal distance from every side.

Angle bisector

P

In Figure 6-11, the angle bisectors intersect at point *P*, which is of equal distance from every side of the triangle.

For any triangle, each side has one median, one perpendicular bisector, and one altitude. Each angle has only one angle bisector.

Interior and Exterior Triangle Details (or Location, Location, Location)

A triangle defines regions relative to itself. How egocentric! Something can be inside a triangle or outside a triangle. And unlike in tennis, if something lands on the line it is counted neither in nor out but rather on the line, a third location. So what's in, what's out, and what's on the line?

On the inside

Everything trapped within the closed space of the triangle's three sides is in the *interior*. Keep in mind that this rule includes all three of the triangle's angles.

The sum measure of the three interior angles of a triangle totals 180° (see Theorem 5-2). Because you're going to be working pretty closely with this information throughout your geometry pursuits, now's a really good time for a proof. Proof 6-1 shows that the sum of the measures of the interior angles of △*XYZ* equals 180°.

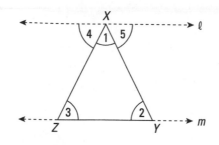

Given: △*XYZ* is a triangle.

Prove: m∠1 + m∠2 + m∠3 = 180°

Before you begin: This is where auxiliary lines come in handy. Draw a line ℓ so that it intersects point *X* and is also parallel to line segment $\overline{ZY}$ of the triangle. Notice that this line creates two additional angles for you to work with — ∠4 and ∠5. You can also extend the line beyond line segment $\overline{ZY}$ so that the steps in the proof are a little clearer. (Doing so isn't necessary. It's just a visual aid.) That line is known collectively as line *m*.

Proof 6-1:
The Sum of Interior Measures of a Triangle Proof

Statements	Reasons
1) m∠4 + m∠1 + m∠5 = 180°	1) A straight angle (line) has a measure of 180° (see Chapter 2).
2) line ℓ ‖ line *m*	2) Through a point not on a line, only one line can be drawn that is parallel to the line (see Chapter 2).
3) m∠2 = m∠5	3) If lines are parallel, alternate interior angles are congruent.
4) m∠3 = m∠4	4) Same as #3.
5) m∠1 + m∠2 + m∠3 = 180°	5) Substitution. In Statement #1, substitute m∠2 for m∠5 and m∠3 for m∠4.

On the outside

Everything outside the closed space of the triangle is in the *exterior*. Now the interior space of a triangle is pretty easy to spot, including the interior angles. But what about exterior angles? Where do they come from? Because a triangle is a polygon, you can treat it as such and extend a side through the vertex to form an exterior angle. Exterior angles are adjacent and supplementary to interior angles (see Figure 6-12).

Figure 6-12:
To form an
exterior
angle (if one
is not
already
drawn for
you), extend
a side
through the
vertex.

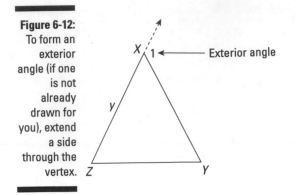

In △*XYZ* in the figure, side *y* (also known as line segment $\overline{ZX}$) has been extended through vertex *X*. This extension forms an angle (labeled 1 for reference) adjacent to ∠*X*. ∠1 is the supplement of ∠*X*. The two remaining angles in △*XYZ* — ∠*Y* and ∠*Z* — are *remote* (nonadjacent) interior angles to ∠1. The sum measure of these two remote interior angles equals the measure of ∠1.

The sum measure of the three exterior angles of a triangle totals 360° (see Theorem 5-6). Which reminds me that now is another really good time for a proof. Proof 6-2 shows that the sum of the measures of the exterior angles of △*XYZ* equals 360°.

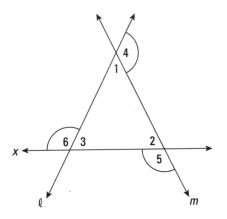

Given: Line *ℓ*, line *m*, and line *x* intersect to form a closed figure with three sides.

Prove: m∠4 + m∠5 + m∠6 = 360°

	Statements		Reasons
	1) Line ℓ, line m, and line x intersect to form a closed figure with three sides.	1)	Given. It's also the definition of a triangle.
	2) $m\angle 1 + m\angle 2 + m\angle 3 = 180°$	2)	The sum of the measures of the interior angles of a triangle equals 180°.
	3) $m\angle 1 + m\angle 4 = 180°$	3)	A straight angle (line) has a measure of 180° (see Chapter 2).
Proof 6-2:	4) $m\angle 2 + m\angle 5 = 180°$	4)	Same as #3.
The Sum of	5) $m\angle 3 + m\angle 6 = 180°$	5)	Same as #3.
Exterior	6) $m\angle 1 + m\angle 4 + m\angle 2 + m\angle 5 + m\angle 3 + m\angle 6 = 540°$	6)	Addition.
Angles of a Triangle	7) $(m\angle 1 + m\angle 4 + m\angle 2 + m\angle 5 + m\angle 3 + m\angle 6) - (m\angle 1 + m\angle 2 + m\angle 3) = 540° - 180°$	7)	Subtraction.
Proof	8) $m\angle 4 + m\angle 5 + m\angle 6 = 360°$	5)	Substitution.

A housekeeping interlude

Figure 6-13 shows the various interior and exterior angles of a triangle, and the following tidbits about the figure tie up the info and relationships about interior and exterior angles into a nice, tidy package (organization counts, my friend):

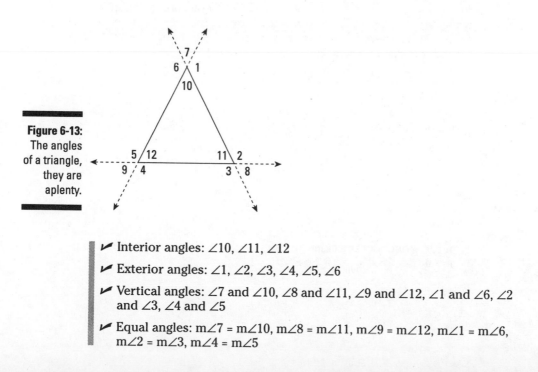

Figure 6-13: The angles of a triangle, they are aplenty.

✔ Interior angles: $\angle 10$, $\angle 11$, $\angle 12$

✔ Exterior angles: $\angle 1$, $\angle 2$, $\angle 3$, $\angle 4$, $\angle 5$, $\angle 6$

✔ Vertical angles: $\angle 7$ and $\angle 10$, $\angle 8$ and $\angle 11$, $\angle 9$ and $\angle 12$, $\angle 1$ and $\angle 6$, $\angle 2$ and $\angle 3$, $\angle 4$ and $\angle 5$

✔ Equal angles: $m\angle 7 = m\angle 10$, $m\angle 8 = m\angle 11$, $m\angle 9 = m\angle 12$, $m\angle 1 = m\angle 6$, $m\angle 2 = m\angle 3$, $m\angle 4 = m\angle 5$

✔ Supplemental angles:

Angles supplemental to ∠10: ∠1, ∠6

   $m\angle 10 + m\angle 1 = 180°$

   $m\angle 10 + m\angle 6 = 180°$

Angles supplemental to ∠11: ∠2, ∠3

   $m\angle 11 + m\angle 2 = 180°$

   $m\angle 11 + m\angle 3 = 180°$

Angles supplemental to ∠12: ∠4, ∠5

   $m\angle 12 + m\angle 4 = 180°$

   $m\angle 12 + m\angle 5 = 180°$

(See Chapter 2 for more on angles.)

On the line

The triangle lives in a world that's divided into an interior and an exterior space. The sides of the triangle itself form the perimeter that's the source of this division. The perimeter is the sum of the measures of the sides that make up a given triangle. If you feel up to adding, finding the perimeter of a triangle is no sweat. Of course, you do have to have the measures of the lengths of all three sides, but then it's as easy as pie. Not to be confused with pi, which is what Chapter 8 is full of. Anyway, the formula for the perimeter of a triangle is $P = a + b + c$, where P is the perimeter and a, b, and c are the lengths of each side of the triangle (see Figure 6-14).

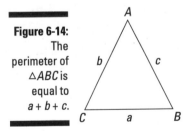

Figure 6-14:
The perimeter of △*ABC* is equal to $a + b + c$.

So if the lengths of the sides are, say, $a = 5$, $b = 6$, and $c = 5$, then the perimeter is equal to $5 + 6 + 5$, or 16. So $P = 16$.

Grouping Triangles by Sides

If you judge triangles (and it's socially OK to judge them — just don't be too harsh) solely by their sides, they can be one of three types: scalene, isosceles, or equilateral.

Scalene

Sounds like a thin fish. A *scalene triangle* has all different-sized parts. None of its sides are congruent in length. For that matter, none of its angles are of the same measure, either. In Figure 6-15, for example, $a \neq b \neq c$.

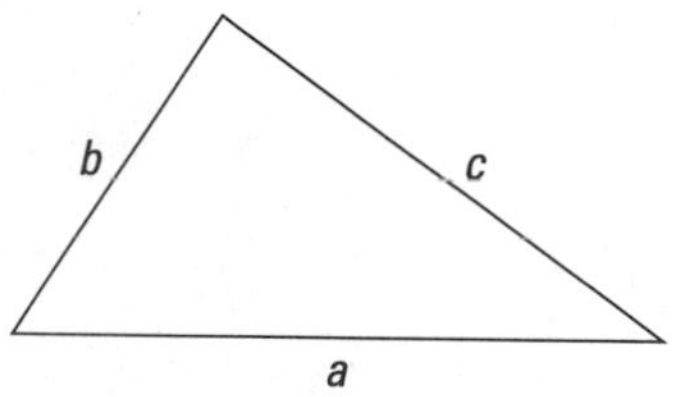

Figure 6-15: A scalene triangle has no congruent parts.

Isosceles

An *isosceles triangle* has two sides of equal length — called *legs*. In Figure 6-16, for example, $b \cong c$. The third side of an isosceles triangle is known as the *base*. The base is the odd man out because it's the nonequal side. The angle opposite the base is the *vertex angle*. It's formed by the two congruent sides, or legs, of the triangle. The vertex angle doesn't contain the base as one of its sides. The angles opposite the legs are called *base angles*. Each of the two base angles is formed by a leg and the base, which is why they're called base angles.

- In Figure 6-16, see those matching slash marks? They indicate the congruent parts of the triangle.

- Typically, the vertex angle is always at the top of an isosceles triangle and the base is at the bottom, but not always. Don't assume.

- An isosceles triangle's legs are of equal length so that the triangle doesn't walk with a limp.

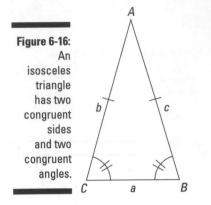

Figure 6-16:
An isosceles triangle has two congruent sides and two congruent angles.

Theorem 6-9: If two sides of a triangle are congruent, then the angles opposite those sides are congruent.

Translation: Sides of a triangle are equal if they're opposite angles of equal measure. This is known as the Base Angles Theorem. You can do a proof on it in the section "Proof city: Proving triangles congruent," later in this chapter.

The Base Angles Theorem — or, more precisely, its converse — is useful in helping you prove that a triangle is isosceles. To get the converse of a statement, you just switch the statement around. Instead of being *if p then q,* the statement is *if q then p* (*p* is the definition of an isosceles triangle). So with this theorem, if you can show that two angles of a triangle are congruent, then the two sides are congruent.

Theorem 6-10: If two angles of a triangle are congruent, then the opposite sides are congruent.

Translation: Angles are equal if they have opposite sides of equal length. This is known as the Converse of the Base Angles Theorem.

In Figure 6-16, $\angle B$ and $\angle C$ are congruent, side *b* and side *c* are congruent.

The converse of an *if . . . then* statement isn't always true. In the case of the Base Angles Theorem, for example, the converse is true, but be careful.

Here's a bit more information about an isosceles triangle that can help you figure out the measures of some of its parts. It involves angle bisectors. They aren't really part of the triangle itself but can add useful information. Because the legs of an isosceles triangle are congruent, the angle bisectors drawn from each leg to its opposite side are congruent. If you draw an angle bisector from the base to the vertex angle of an isosceles triangle, the angle bisector divides that triangle into two congruent triangles (see Figure 6-17). (For more info about congruent triangles, check out the last section of this chapter.)

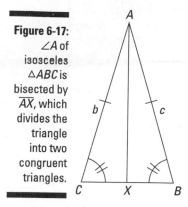

Figure 6-17:
∠*A* of isosceles △*ABC* is bisected by $\overline{AX}$, which divides the triangle into two congruent triangles.

Theorem 6-11: The bisector of the vertex angle of an isosceles triangle divides the triangle into two congruent triangles.

Translation: When an angle bisector is drawn to the vertex angle of an isosceles triangle (the angle opposite the nonequal side), then the angle bisector divides the triangle into two congruent triangles.

In Figure 6-17, △*CAX* and △*BAX* are congruent.

Corollary 6-1: An angle bisector of the vertex angle of an isosceles triangle is a perpendicular bisector of the base of the triangle.

Translation: In an isosceles triangle, when an angle bisector is drawn from the vertex angle to the opposite side of the angle, not only does the angle bisector meet the side at a 90° angle, but it also splits that side into two equal segments.

In Figure 6-17, $\overline{AX}$ bisects ∠A and is also perpendicular to $\overline{CB}$.

Corollary 6-2: The bisector of the vertex angle of an isosceles triangle bisects the base.

Translation: This one should probably go before Corollary 6-1 because it says less, but see whether you can find the difference. Basically, in an isosceles triangle, when an angle bisector is drawn from an angle to the opposite side of the angle, the angle bisector splits that side into two equal segments.

In Figure 6-17, $\overline{AX}$ bisects $\overline{CB}$.

Equilateral

Equilateral means "equal sides." So it follows that an *equilateral triangle* has equal sides — three equal sides, to be more precise. In Figure 6-18, for example, $a = b = c$. In a triangle, if the sides are congruent, then the opposite angles are congruent. An isosceles triangle has two congruent sides with two congruent angles. An equilateral triangle follows the same pattern. It has three congruent sides and, as logic would lead you to believe, three congruent angles.

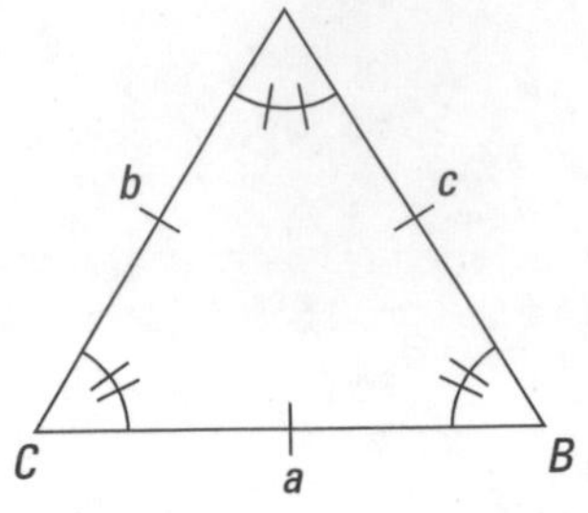

Figure 6-18:
In an equilateral triangle, all sides and all angles are congruent.

An equilateral triangle is also an isosceles triangle because an isosceles triangle has at least two equal sides. An equilateral triangle has three congruent sides, so it qualifies.

The sum of the measures of the interior angles of a triangle is 180°. If a triangle has three angles, and each angle is of equal measure, then the measure of each angle is 180°/3, or 60°.

Corollary 6-3: An equilateral triangle is also equiangular.

Corollary 6-4: The three angles of an equilateral triangle each have a measure of 60° (see Theorem 5-2 and Proof 6-1).

Translations: If a triangle has three equal sides, then it has three equal angles. Because all the angles are equal, and you know that the total measure of the interior angles of a triangle is 180°, one of those interior angles has a measure of 60°.

In Figure 6-18, $a = b = c$ and $\angle A = \angle B = \angle C$.

An equilateral triangle's altitude, perpendicular bisector, and angle bisector are the same line. You can determine the length of this line by multiplying the length of any side by $\frac{\sqrt{3}}{2}$ (see Figure 6-19).

Theorem 6-12: The altitude of an equilateral triangle is the square root of three divided by two multiplied by the length of a side of the triangle.

Translation: This one's pretty self-explanatory, but you need the length of a side of the triangle for it to be of any use.

In Figure 6-19, $\overline{AZ}$ is the altitude of triangle ABC. The length of $\overline{AZ}$ is the length of a side times $\frac{\sqrt{3}}{2}$ The length of c is given as 6. From this info, you can determine that the length of $\overline{AZ}$ is $\frac{\sqrt{3}}{2}$, or approximately 5.20 units.

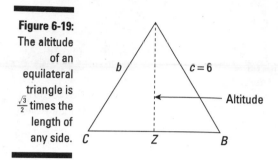

Putting it all together

I like to keep everything straight, so I love flowcharts. If you do, too, check out Figure 6-20, which shows how to determine the type of triangle from its sides.

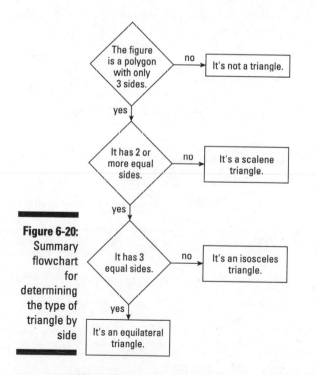

Grouping Triangles by Angles

Just as you can group triangles by their sides, you can group them by the measurements of their angles. The groups include acute, obtuse, equiangular, and right.

Acute as a baby's bottom

An *acute triangle* is one whose angles are all less than 90° — not the total measure of all but for each. Form an L with your thumb and forefinger and then move your forefinger down toward your thumb until it touches the thumb and your fingers are pinched together. That's the range of degrees of any given angle in an acute triangle — between (but not including) 0° and 90° (see Figure 6-21).

Figure 6-21:
An acute
triangle has
three angles
that each
measure
less
than 90°.

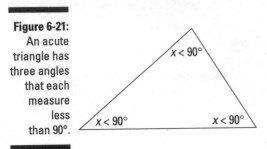

As is usual with geometry stuff, a formula is available for determining whether a triangle is acute. But before you get into the formula details, you may need a quickie dose of some squares. Table 6-1 gives you the squares of some frequently used numbers.

Table 6-1:	A Cheat Sheet on Squares
The Number (n)	*The Square of the Number (n^2)*
3	9
4	16
5	25
6	36
7	49
8	64
9	81
10	100
11	121
12	144

You can determine whether a triangle is acute without measuring its angles, but you have to know the measures of all the sides. You can say that you have an acute triangle if the square of the length of the longest side is less than the sum of the squares of the lengths of the other two, shorter sides. So the formula is $c^2 < a^2 + b^2$.

In Figure 6-22, side a is 4-feet long, side b is 4-feet long, and side c is 5-feet long, so the longest side is 5 feet. If the triangle in Figure 6-22 is acute, then the sum of the squares of the two 4-foot sides should be greater than the square of the 5-foot side:

Figure 6-22:
For a triangle to be acute, the sum of the squares of the two shorter sides is greater than the square of the longest side.

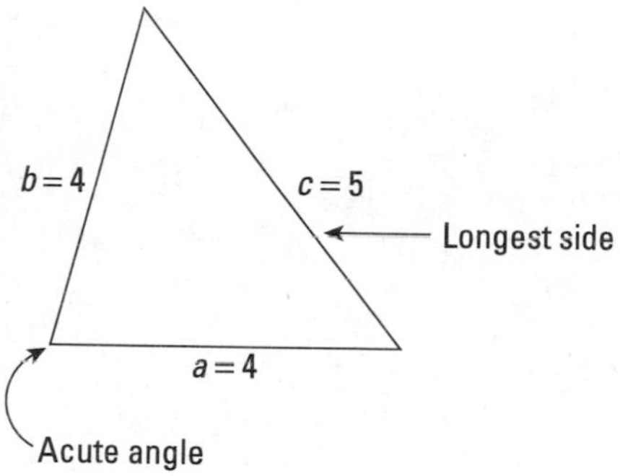

$$(4)^2 + (4)^2 \stackrel{?}{=} (5)^2$$
$$16 + 16 \stackrel{?}{=} 25$$
$$32 > 25$$

Because 32 (the sum of the squares of the shorter sides) is greater than 25 (the square of the longest side), the triangle is acute. Notice that two of the sides are equal. This triangle is not only acute, but it's also an isosceles triangle.

Theorem 6-13: If the square of the length of the longest side is less than the sum of the squares of the lengths of the other two shorter sides, then the triangle is acute.

Translation: Use the lengths of the sides and properly plug them into the formula. If the logic of the equation holds, then the triangle is acute. If the logic doesn't hold, then the triangle is some other kind of triangle.

Obtuse but certainly not dull

An *obtuse triangle* is a triangle with *one* obtuse angle. Think about this: An obtuse angle has a measure of greater than 90°, and the sum of the interior angles of a triangle equals 180°. Put those two tidbits together, and you see why an obtuse triangle can have only one obtuse angle.

If you don't have your protractor handy, you can determine whether a triangle is obtuse by using the measures of its sides or, more precisely, the squares of those sides. If the square of the length of the longest side is greater than the sum of the squares of the other two sides, then you have a triangle that is obtuse. In Figure 6-23, $a = 8$, $b = 3$, and $c = 10$. Apply the formula to see how it works:

Figure 6-23:
For a triangle to be obtuse, the sum of the squares of the two shorter sides is less than the square of the longest side.

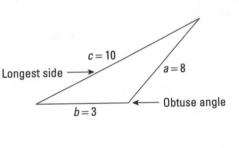

$$(8)^2 + (3)^2 \overset{?}{=} (10)^2$$

$$64 + 9 \overset{?}{=} 100$$

$$73 < 100$$

Theorem 6-14: If the square of the length of the longest side is greater than the sum of the squares of the lengths of the other two, shorter sides, then the triangle is obtuse.

Translation: Use the lengths of the sides and properly plug them into the formula. If the logic of the equation holds, then the triangle is obtuse.

Theorem 6-15: A triangle may have at most one obtuse angle.

Translation: This one can actually be categorized as a corollary of Theorem 5-2, the Angle Sum Theorem of a Triangle. A triangle has 180° to be shared among three angles. If one angle hogs more than 90° just for itself, that leaves less than 90° to be split between the remaining two angles. By definition, that makes the remaining two angles acute.

Equiangular: It's all the same to me

An *equiangular triangle* is a triangle with three equal angles. Incidentally, it has three equal sides, too. All sides equal, all angles equal. Oh, and each angle has a measure of 60°. That's about it. In Figure 6-24, $\angle A = \angle B = \angle C$.

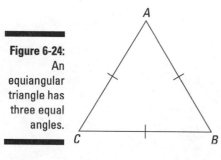

Figure 6-24:
An equiangular triangle has three equal angles.

Send in the auxiliary. The median, altitude, perpendicular bisector, and angle bisector are one and the same in an equiangular triangle. An equiangular triangle is also equilateral. What goes for one goes for the other.

Right as a 90-degree summer's day

A *right triangle* has a right angle as one of its angles. The other two angles are acute because the sum of three angles in a triangle must equal 180°. The two angles are also complementary; if a right angle equals 90°, that leaves only 90° more to be split between the remaining two angles of the triangle. If the two acute angles are each 45° in measure, then the right triangle is also an isosceles triangle or, more precisely, a right isosceles triangle.

Corollary 6-5: A triangle may have at most one right angle (see Theorem 5-2 and Proof 6-1).

Translation: A triangle has three angles with a total measure of 180°. If one angle is a right angle, it has a measure of 90°, leaving only 90° to be split between two angles.

Corollary 6-6: The acute angles of a right triangle are complementary (see Theorem 5-2 and Proof 6-1).

Translation: A triangle has three angles with a total measure of 180°. If one angle is a right angle, it has a measure of 90°. The remaining 90° is split between two angles. The definition of *complementary* is that the sum of two angles equals 90°.

Time for a proof. Proof 6-3 shows that a right triangle contains one right interior angle and two complementary angles.

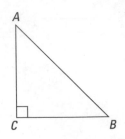

Given: △ABC is a right triangle, and ∠C = 90°.

Prove: ∠A + ∠B are complementary

Proof 6-3:
The Right
Triangle
Proof

Statements	Reasons
1) △ABC is a right triangle	1) Given.
2) ∠A + ∠B + ∠C = 180°	2) The sum of the measure of the interior angles of a triangle equals 180°.
3) ∠C = 90°	3) Given.
4) ∠A + ∠B + 90° = 180°	4) Substitution (see Chapter 3).
5) ∠A + ∠B = 90°	5) Angle Addition (see Chapter 3).
6) ∠A + ∠B are complementary	6) The definition of complementary angles. Two angles whose sume measures equal 90° are complementary (see Chapter 2).

When an altitude is drawn from the right angle to the hypotenuse it creates two triangles that are similar to each other. I will address similarity further in Chapter 10, so for now you are just going to have to take my word for it. Should the parent triangle just happen to be a right isosceles, the altitude drawn to the hypotenuse pulls double-duty as it is also the median. Another interesting twist exists: The hypotenuse of both of the newly created triangles is the altitude of their parent triangle. Also, because the altitude and the median are one and the same, the length of that segment (regardless of what you call it) is equal to one-half the length of the hypotenuse. Yes indeedy. That's a bit hard to follow. So just check out Figure 6-25.

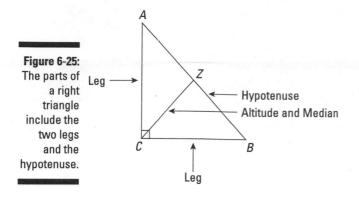

Figure 6-25:
The parts of a right triangle include the two legs and the hypotenuse.

In Figure 6-25, the length of the median (or altitude) $\overline{CZ}$ is one-half the length of the hypotenuse $\overline{AB}$ because $\triangle ABC$ is a right isosceles triangle.

Theorem 6-16: The length of the median to the hypotenuse of a right isosceles triangle is equal to one-half the length of the hypotenuse.

Translation: If you have a right isosceles triangle, the length of the line from the right angle to the midpoint of the hypotenuse (the side opposite the right angle) is one-half the length of the hypotenuse.

The appendages of a right triangle have a special relationship to the hypotenuse. The sum of the squares of the two legs equals the square of the hypotenuse. This is known as the Pythagorean Theorem, and it's written in geometric shorthand as $a^2 + b^2 = c^2$.

Theorem 6-17: For a right triangle, the sum of the squares of the lengths of the legs equals the length of the square of the hypotenuse. That is, $a^2 + b^2 = c^2$, which is otherwise known as the Pythagorean Theorem.

Pythagoras *loved* order

The works of Pythagoras, a Greek mathematician circa 532 B.C., attempted to impose order on the universe. He believed that the universe had a formal structure that could be expressed in numbers and proportions. Pythagoras is believed to have solved the first proof showing the relationship between the sides of a right triangle. Pythagoras showed that if a right triangle was reproduced several times it would form a square. From this info, he developed a statement regarding the order of things. That statement is famous in geeky geometry circles and is known as the Pythagorean Theorem. It shows that the square of the length of the side opposite the right angle is equal to the sum of the squares of the other two sides: $c^2 = a^2 + b^2$. Pythagoras also proved that the sum measure of the angles of a triangle equals that of two right angles.

Translation: The assumption here is that the triangle is a right triangle and that when the lengths of the sides are properly popped into the equation, the logic of the equation holds.

Turning things around and looking at the Converse of the Pythagorean Theorem can help you prove that a triangle is a right triangle. You do need the measures of all the sides, but no angle measurements are necessary. Plug the measures of the sides into the formula $a^2 + b^2 = c^2$, with c as the longest length. If the logic holds once you've calculated the numeric values, then you can conclude that the triangle is a right triangle (see Figure 6-26).

Figure 6-26:
For a triangle to be a right triangle, then $a^2 + b^2 = c^2$.

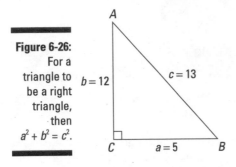

In Figure 6-26, the lengths of the sides are $a = 5$, $b = 12$, and $c = 13$. If $\triangle ABC$ is really a right triangle, then if you plug the information into the formula, the logic will hold up. That means $5^2 + 12^2 = 13^2$. Yep. Because 169 = 169, then $\triangle ABC$ is a right triangle.

Theorem 6-18: If the square of the length of the longest side equals the sum of the squares of the lengths of the other two shorter sides, then the triangle is a right triangle. That is, if $c^2 = a^2 + b^2$, then you can conclude that the triangle is a right triangle. This is the Converse of the Pythagorean Theorem.

Translation: This is just another way of looking at Theorem 6-17. Instead of assuming that the triangle has a right angle, plug the side lengths into the formula, and if the logic holds true, then you can conclude that the triangle is a right triangle.

No matter what you've heard, a relationship is never 50-50. The relationships within a right triangle are no exception. Some special angle relationships exist. You know that the measure of one angle of a right triangle is 90°. But what about the other two angles? You know that they're both acute, but what's the rest of the story?

30-60-90 relationships

When the acute angles in a right triangle have measures of 30° and 60°, the following order is imposed on the universe (see Figure 6-27 as you review each of the following relationships):

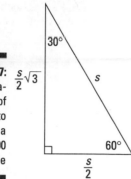

✔ The length of the leg opposite the 30° angle is equal to one-half the length of the hypotenuse. In geometric shorthand, it's $a = \frac{s}{2}$.

✔ The length of the leg opposite the 60° angle is equal to one-half the length of the hypotenuse multiplied by $\sqrt{3}$: In geometric shorthand, it's $b = \frac{s}{2}\sqrt{3}$.

✔ The length of the side opposite the 60° angle is equal to the length of the side opposite the 30° angle multiplied by $\sqrt{3}$. In geometric shorthand, it's $b = a\sqrt{3}$.

45-45-90 relationships

This is as close to even input as you're going to get. This is a right isosceles triangle. You may notice that some of the relationships look familiar (see Figure 6-28):

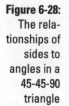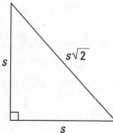

✔ The lengths of the legs are equal. In geometric shorthand, it's $a = b$.

✔ The length of the hypotenuse is equal to the length of the leg multiplied by $\sqrt{2}$. In geometric shorthand, it's $c = a\sqrt{2}$ or $c = b\sqrt{2}$.

✔ The length of the leg is equal to one-half the length of the hypotenuse multiplied by $\sqrt{2}$. In geometric shorthand, it's $a = 1/2c\sqrt{2}$ or $b = 1/2c\sqrt{2}$.

To get the length of the hypotenuse from the length of the side, because each side length is s, then the length of the hypotenuse is $s\sqrt{2}$, where s is the length of a leg of the right isosceles triangle.

To get the length of the side when you have the length of the hypotenuse, multiply one-half the length of the hypotenuse by $\sqrt{2}$, or $1/2c\sqrt{2}$.

Putting it all together

Because there's so much to know about this three-sided figure affectionately known as the triangle, I've put together Figure 6-29, which is a handy-dandy summary chart for determining a triangle's type by its angles.

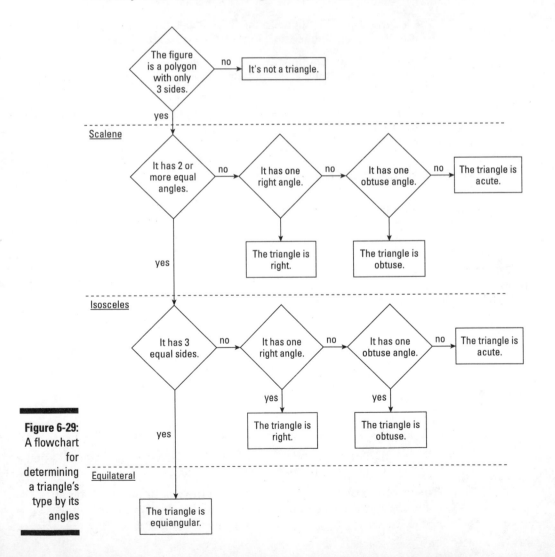

Figure 6-29:
A flowchart for determining a triangle's type by its angles

The Triangle Construction Zone

This section shows you how to draw a triangle of your own. You need something to draw with. Crayons are good. Just kidding. A pen or, better yet, a pencil is probably a bit more age appropriate. You also need a protractor and a bit of information about the triangle; the lengths of the three sides and the measures of all three angles are optimal.

Look over the information that you're given. See whether anything gives an indication of what your triangle will look like when it's completed. For example, suppose that you're given information for three equal sides and three equal angles. The length of the sides are $\overline{AB}$ = 3 cm, $\overline{BC}$ = 3 cm, and $\overline{CA}$ = 3 cm, and the measures of the angles are $\angle A$ = 60, $\angle B$ = 60, and $\angle C$ = 60. Once your triangle is completed, it'll be an equilateral and equiangular triangle (see Figure 6-30).

Figure 6-30:
To construct this triangle, use a protractor and the information given about the lengths of its sides and the measures of its angles.

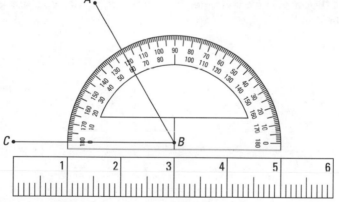

To draw the triangle shown in Figure 6-30

1. **Draw a 3 cm line that represents segment $\overline{BC}$.**

2. **Get out your protractor and mark off the measure of $\angle B$ by placing a dot at 60°.**

3. **Use your ruler to draw a 3 cm line from point B through the dot you just placed at 60°.**

   The top of this line is point A.

4. **Using your protractor, use the line $\overline{BC}$ that you just drew as the 0° line and mark off another angle of 60°.**

   The only thing left to do is close the triangle.

5. **Use your ruler to draw a straight line through the 60° angle mark you just made and extend the line from point A to point C.**

How often do you get exactly what you want? When the information given is less than optimal, it must at least be sufficient. By sufficient, I mean that you need at least enough information to draw the triangle without having to guess about the measure of an angle or side to complete the figure. For example, in Figure 6-30, if the measurements of $\angle A$ and $\angle C$ were missing, you could look at the lengths of the sides. You know that an equilateral triangle is equiangular, so look no further. The measurements of both $\angle A$ and $\angle C$ are 60°.

Calculating the Area of a Triangle

When three lines come together to form a triangle, these lines enclose some space within them. You can measure this interior region of space just as you measure the sides or angles of a triangle. A side is measured in distance, an angle is measured in degrees, and space is measured in area. To find the area of a triangle, you can use the generic formula $A = \frac{1}{2}bh$, where A is the area, b is the length of the base, and h is the height, or altitude, to that side. In Figure 6-31, for example, $A = \frac{1}{2}(8)(3)$, or 12 square units.

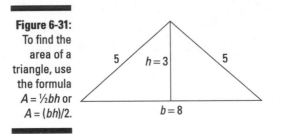

Figure 6-31:
To find the area of a triangle, use the formula $A = \frac{1}{2}bh$ or $A = (bh)/2$.

Theorem 6-19: The area of a triangle is $A = \frac{1}{2}bh$.

Translation: What can I say? It's a formula. Plug in the numbers. Can't use this one, though, if you don't have the info about the length of the base or the height (altitude) to that side.

If you draw a median in a triangle, it divides the triangle into two triangles with equal areas. How is this so? First, the median goes from the vertex to the midpoint of the opposite side. Apply the formula for the area of a triangle to each of the triangles formed by the median, and you see what happens. The median divides the base into two equal segments, and because the altitude is the shared median, the areas of the two triangles are congruent (see Figure 6-32).

Designer formulae are available for finding the area of a right triangle and an equilateral triangle. The area for a right triangle is $A = \frac{1}{2}ab$, where a and b are the measures of the legs (see Figure 6-33). The area for an equilateral triangle is $A = (s^2/4)\sqrt{3}$ (see Figure 6-34).

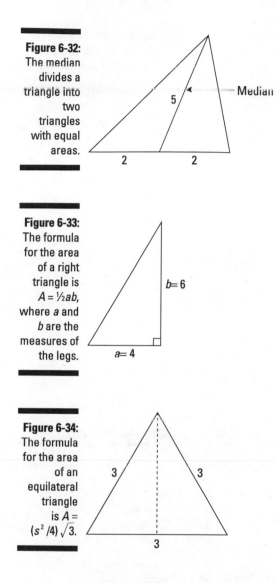

Figure 6-32:
The median divides a triangle into two triangles with equal areas.

Figure 6-33:
The formula for the area of a right triangle is $A = \frac{1}{2}ab$, where a and b are the measures of the legs.

Figure 6-34:
The formula for the area of an equilateral triangle is $A = (s^2/4)\sqrt{3}$.

In Figure 6-33, $a = 4$ and $b = 6$. $A = \frac{1}{2}(4)(6)$, which boils down to 12 square units. In Figure 6-34, the length of side s is 3. So $A = (3^2/4)\sqrt{3}$, which is about 3.9 square units.

Of course, if you don't like any of those formulae, there's always another one, but you need the lengths of all the sides to use it. You can find the area of a triangle by using the formula $A = \sqrt{s(s-a)(s-b)(s-c)}$, where s is half the perimeter of the triangle — $s = \frac{1}{2}(a + b + c)$ and a, b, and c are the lengths of the sides of the triangle. This formula is sometimes referred to as Heron's Formula.

To calculate the area of the triangle shown in Figure 6-35, you first need to take care of some business. You need half the perimeter of a triangle so that you have a value to plug in for s. If $s = \frac{1}{2}(a + b + c)$, then $s = \frac{1}{2}(5 + 6 + 5)$. After everything is condensed and simplified, s equals 8.

The lengths of sides a, b, and c are 5, 6, and 5, respectively. Put this info into the formula and you get $A = \sqrt{8(8-5)(8-6)(8-5)}$ or $\sqrt{8(3)(2)(3)}$ or, better yet, just 12.

Figure 6-35:
If you've got the lengths of the sides, you can find the area by using Heron's Formula.

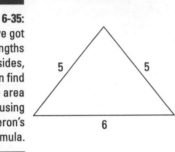

Want an all-in-one-spot summary about the sides, angles, and areas of a triangle? You got it. Check out Table 6-2.

Table 6-2:		An All-in-One Summary about the Triangle				
	Scalene	*Isosceles*	*Equilateral/ Equiangular*	*Acute*	*Right*	*Obtuse*
Sides	No sides are equal.	Two sides are equal.	Three sides are equal.	Possibility of two or three equal sides.	Possibility of two equal sides.	Possibility of two equal sides.
Angles	No angles are equal.	Two angles are equal.	Three angles are equal.	All angles are acute; possibility of two or three equal angles.	One right angle; possibility of the other two angles being congruent.	One obtuse angle; possibility of the other two acute angles being congruent.
Area	$A = \frac{1}{2}bh$	$A = \frac{1}{2}bh$	$A = (s^2/4)\sqrt{3}$	$A = \frac{1}{2}bh$	$A = \frac{1}{2}ab$	$A = \frac{1}{2}bh$

Looking at Congruent Triangles (They're Sooooo Conformist)

Two triangles are *congruent* to each other if they are of the exact same shape and size. Think of congruent triangles as clones of each other. To have an exact copy, you need to zero in on the core genetic information that defines the very essence of a triangle. Triangles are defined by their sides and by their angles, making that a good place to start. Look at both triangles; systematically mark them up: Mark the same number of dashes on congruent sides and on congruent angles and put a square inside a right angle. You're looking for a one-to-one correspondence between angles and sides. Correspondence doesn't have to do with communication but with pairing (see Figure 6-36).

Figure 6-36:
Marking up
△*WLA* and
△*TNC* —
looking for
corres-
pondence
between
angles and
sides

For △*WLA* and △*TNC*, three pairs of corresponding sides and three pairs of corresponding angles exist. The symbol for correspondence is ←→.

- ✔ Correspondence of angles: ∠*W* ←→ ∠*T*, ∠*L* ←→ ∠*N*, and ∠*A* ←→ ∠*C*.
- ✔ Correspondence of sides: $\overline{WL}$ ←→ $\overline{TN}$, $\overline{LA}$ ←→ $\overline{NC}$, and $\overline{AW}$ ←→ $\overline{CT}$.

When looking at correspondence of triangles, consider that △*WLA* and △*TNC* are different from △*AWL* and △*NTC*. And which parts are matched changes as a result.

Congruent run-of-the-mill triangles

Seems like it could be a lot of work to prove triangles congruent — especially if you have to go through all six angles and all six sides. Lucky for you, short-cuts are available. That is, you don't have to show *all* corresponding parts congruent. Only certain ones. Of course, the location of the congruent parts in the triangle also makes a difference. Remember, this is geometry, and you have to play by the rules.

Two of the shortcuts are sandwiches. You can show a sandwich of congruent parts. You can have an angle sandwich or a side sandwich. A sandwich is named for what's in between. If you have an angle sandwich, for example, then the angle is in between two sides. The angle is considered to be *included* between the two sides. This method of showing congruence is called *side-angle-side,* or *SAS* (see Figure 6-37).

Figure 6-37:
Showing
congruent
triangles
using
the side-
angle-side
(SAS)
method

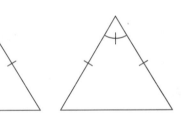

If you have a side sandwich, a side is included between two angles (see Figure 6-38). This method of showing congruence is called *angle-side-angle (ASA).* Incidentally, the side sandwich comes with soup.

Figure 6-38:
Showing
congruent
triangles
using the
angle-side-
angle (ASA)
method

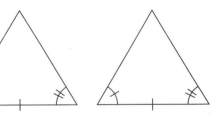

Because you can't always get a sandwich when you want one, you can find congruence between triangles in two other ways. One is by proving that all three sides of one triangle are congruent to all three sides of the other triangle (see Figure 6-39). This method is known as congruence by *side-side-side (SSS)*.

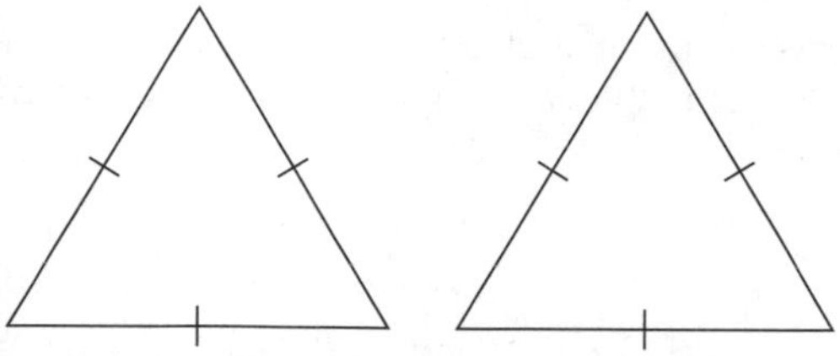

Figure 6-39:
Showing
congruent
triangles
using the
side-side-
side (SSS)
method

REMEMBER

Equilateral triangles are equiangular.

The other method is to show that two consecutive angles and the side that is not between the angles are congruent to corresponding parts in another triangle (see Figure 6-40). This method is known as congruence by *angle-angle-side (AAS)* or *side-angle-angle (SAA)*.

Figure 6-40:
Showing
congruent
triangles
using the
angle-
angle-side
(AAS) or
side-angle-
angle (SAA)
method

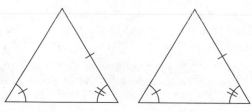

In general, to show that two triangles are congruent, you can consult the following list. You only have to show that one of the following statements (or postulates) applies:

- *Postulate 6-1:* Two sides and the included angle are congruent to corresponding parts of the other triangle ($SAS \cong SAS$).

- *Postulate 6-2:* Two angles and the included side are congruent to corresponding parts of the other triangle ($ASA \cong ASA$).

- *Postulate 6-3:* The three sides are congruent to corresponding sides of the other triangle ($SSS \cong SSS$).

- *Postulate 6-4:* Two consecutive angles and the nonincluded side are congruent to corresponding parts of the other triangle ($AAS \cong AAS$ or $SAA \cong SAA$).

Congruent right triangles

Right triangles are so special. They can actually be easier to work with than other triangles because you already have some information about them.

One angle of a right triangle is a right angle, and the other two angles are acute and complementary.

With other triangles, you need to show that three corresponding parts are congruent, but because there are some givens with a right triangle, you only need two. Five ways (postulates) are available for you to prove congruent right triangles:

- *Postulate 6-5:* Two right triangles are congruent if the lengths of the two legs of one triangle are congruent to the legs of the other triangle (leg-leg).

- *Postulate 6-6:* Two right triangles are congruent if the corresponding leg and hypotenuse of one triangle are congruent to those of the other triangle (hypotenuse-leg).

- *Postulate 6-7:* Two right triangles are congruent if an acute angle of one triangle is congruent to the corresponding acute angle of the other triangle and the hypotenuses are the same length (hypotenuse-angle).

- *Postulate 6-8:* Two right triangles are congruent if an acute angle and its adjacent leg of one triangle are congruent to the corresponding parts of the other triangle (adjacent leg-angle).

- *Postulate 6-9:* Two right triangles are congruent if an acute angle and its opposite leg of one triangle are congruent to the corresponding parts of the other triangle (opposite leg-angle).

Congruent triangles and corresponding parts

Once you've determined that a triangle is congruent to another triangle, it makes sense that the two congruent triangles have the same lengths for three corresponding sides and the same measures for three corresponding angles. You've shown this even without the measure of every angle and every side of the triangles — just the critical ones.

Believe it or not, you can now make a statement. If two triangles are proven congruent, then the corresponding parts of these triangles are congruent (CPCTC). In other words, if you show that two triangles are congruent, you can state that any corresponding parts of the two triangles are congruent. May not seem like a big deal now, but it is. Trust me.

CPCTC = Corresponding Parts of Congruent Triangles are Congruent.

Proof city: Proving triangles congruent

The best way to get good at proofs is to just do them. Lots of them. In this chapter, I introduce quite a few theorems without proving them. In this section, I walk you through some of them.

Before you get started on the proofs, though, I want to give you a quickie brain dump — with a twist — of some rules that you may need to use to show congruence:

- **Reflexive property:** Any geometric figure is congruent to itself ($\triangle TNC \cong \triangle TNC$).

- **Symmetric property:** A congruence may be reversed (if $\triangle TNC \cong \triangle XYZ$, then $\triangle XYZ \cong \triangle TNC$).

- **Transitive property:** Two geometric figures congruent to the same geometric figure are congruent to each other (if $\triangle XYZ \cong \triangle TNC$ and $\triangle TNC \cong \triangle ABC$, then $\triangle XYZ \cong \triangle ABC$).

In Proof 6-4, to show that the two sides of $\triangle ABC$ are congruent, you need to prove that opposite angles are congruent. This is the proof for the Base Angles Theorem (see Theorem 6-9).

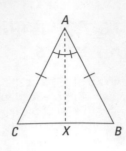

Given: $\triangle ABC, \overline{AB} \cong \overline{BC}$

Prove: $\angle B \cong \angle C$

Before you begin: Draw an angle bisector from vertex A to point X.

	Statements	Reasons
Proof 6-4: The Congruent Opposite Angles Proof	1) $\overline{AB} \cong \overline{BC}$	1) Given.
	2) $\overline{AX}$ bisects $\angle CAB$	2) An angle has only one bisector.
	3) $\overline{AX} \cong \overline{AX}$	3) Reflexive.
	4) $\triangle CAX \cong \triangle BAX$	4) SAS (side-angle-side).
	5) $\angle B \cong \angle C$	5) CPCTC (corresponding parts of congruent triangles are congruent).

In Proof 6-5, to show that the two angles of $\triangle ABC$ are congruent, you need to prove that opposite sides are congruent. This is the proof for the Converse of the Base Angles Theorem (see Theorem 6-10).

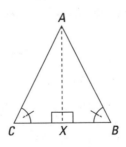

Given: $\triangle ABC, \angle B \cong \angle C$

Prove: $\overline{AB} \cong \overline{BC}$

Before you begin: Draw a perpendicular bisector to $\overline{CB}$ at point X.

Statements	Reasons
1) $\angle B \cong \angle C$	1) Given.
2) $\overline{AX}$ bisects $\overline{CB}$ at point X.	2) A line only has one bisector
3) $\overline{AX}$ is perpendicular to $\overline{CB}$ at point X.	3) A perpendicular bisector intersects a line at a 90° angle.
4) $\angle BXA = 90°$	4) Perpendicular angles equal 90°.
5) $\angle AXC = 90°$	5) Same as #4.
6) $\angle BXA = \angle AXC$	6) Substitution (see Chapter 2).
7) $\overline{AX} \cong \overline{AX}$	7) Reflexive.
8) $\triangle BXA \cong \triangle AXC$	8) AAS $\cong$ AAS.
9) $\overline{AB} \cong \overline{BC}$	9) CPCTC (corresponding parts of congruent triangles are congruent).

Proof 6-5:
The Congruent Opposite Sides Proof

In Proof 6-6, you need to use the figure in Proof 6-5 to prove that $\overline{AB} \cong \overline{BC}$ by using the angle-opposite leg method for proving congruent right triangles. Doing so saves a step. Keep the same Given, Prove, and Before You Begin statements that are used in Proof 6-5 (see Theorem 6-10).

Statements	Reasons
1) $\angle B \cong \angle C$	1) Given.
2) $\overline{AX}$ bisects $\overline{CB}$ at point X.	2) A line only has one bisector
3) $\overline{AX}$ is perpendicular to $\overline{CD}$ at point X.	3) A perpendicular bisector intersects a line at a 90° angle.
4) $\angle BXA = 90°$	4) Perpendicular angles equal 90°.
5) $\angle AXC = 90°$	5) Same as #4.
6) Omit this step for this proof (see #8 from Proof 6-5): $\angle BXA = \angle AXC$	6) Omit this step for this proof (see #8 from Proof 6-5): Substitution (see Chapter 2).
7) $\overline{AX} \cong \overline{AX}$	7) Reflexive.
8) $\triangle BXA \cong \triangle AXC$	8) Angle-opposite leg $\cong$ Angle-opposite leg.
9) $\overline{AB} \cong \overline{BC}$	9) CPCTC (corresponding parts of congruent triangles are congruent).

Proof 6-6:
The Congruent Angle, Opposite-Leg Right Triangle Proof

Proof 6-7 is the proof for Theorem 6-11. The bisector of the vertex angle of an isosceles triangle divides the triangle into two congruent triangles.

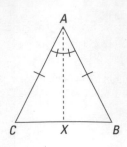

Given: △ABC is an isosceles triangle, and A is the vertex.

Prove: △ACX ≅ △ABX

Before you begin: Add an angle bisector from angle A to point X.

Statements	Reasons
1) △ABC is an isosceles triangle	1) Given.
2) $\overline{AX}$ bisects ∠A into equal angles	2) Definition of a bisector. An angle bisector divides an angle into two equal angles (see Chapter 2).
3) ∠BXA ≅ ∠AXC	3) A bisector of the vertex angle of an isosceles is a perpendicular bisector to the base.
4) $\overline{AC} = \overline{AB}$	4) $\overline{AC}$ and $\overline{AB}$ are sides of an isosceles triangle.
5) $\overline{AX} ≅ \overline{AX}$	5) Reflexive.
6) △ACX ≅ △ABX	6) SAS ≅ SAS.

Proof 6-7:
The
Isosceles
Bisector
Proof

Chapter 7

How to Develop Your Quads

· ·

In This Chapter

▶ The properties of the average quadrilateral

▶ Details and more details about parallelograms, rectangles, rhombi, squares, and trapezoids

▶ The scoop on kites and arrowheads (yep, they're quads, too)

· ·

Set the amount of weight you want to lift. Sit in the leg extension machine with your ankles behind the lifting arm and your hands on the grips at your side. With your legs bent at 90°, extend them forward until they're almost straight. Be careful not to lock your knees. Hold for a count of three and then slowly lower your legs back to the starting position without allowing the weights to touch the stack. You want to keep constant tension on the muscle. Do this exercise a couple times a week, and you'll have incredible quads.

That's about it for the physical quad development tips for this chapter. From here on out, it's strictly geometry. You get to explore all those extraordinary properties of quadrilaterals. Even more special are the subgroupings of quads. They include parallelograms, rectangles, rhombi, squares, and trapezoids.

The Properties of Quadrilaterals

The most identifiable quality of a quadrilateral comes from its name — *quadri,* meaning "four," and *lateral,* meaning "side." A *quadrilateral* is a polygon that has four sides. In Figure 7-1, line segment $\overline{AB}$ leads to line segment $\overline{BC}$, which leads to segment $\overline{CD}$, which leads to segment $\overline{DA}$ to close the figure. That's all the sides, so I don't need to lead you any further.

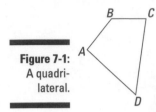

Figure 7-1:
A quadri-
lateral.

When two sides are right after each other as you travel around a polygon, they're called *consecutive sides.* Two sides that are *not* next to each other but that sit across from each other are known as *opposite sides.* Consecutive sides have a common point of intersection. This common point is known as the *vertex.* Every quadrilateral has four vertices. Vertices at either end of any given line segment are known as the *consecutive angles.* You name a quadrilateral by using the four consecutive vertices. Which vertex you start with or which way you go around the figure doesn't matter. Just pick a vertex and a direction and then continue around the figure until you've used up all the letters of the vertices in the name. No repeats allowed.

Taking shape

Because the number of sides of a quadrilateral is always four, the shape of a quadrilateral is altered by the lengths of its sides and the degree measures at which these sides meet. It is important when working with quads to first determine which of two groups the quadrilateral belongs to. Is it convex or concave? If a quad is *concave,* then the measure of one or more of its angles has a measure of greater than 180°. Look for angles that bend farther than a straight line. Those are the angles that are greater then 180° and make a figure concave. If you're still in doubt, yet another way is available for determining concavity without measuring. You can draw a line from the middle of one side to the middle of another. If the line travels through the outside, or exterior, region of the quad, then the quad is concave. With a convex figure, a line drawn from the midpoint of one segment to another segment travels inside, or interior to, the polygon, and all measures of the interior angles are between 0° and 180°. In most cases in this book (as in most geometry books), I refer to convex quads (unless I say otherwise). Figure 7-2 shows examples of both convex and concave quads.

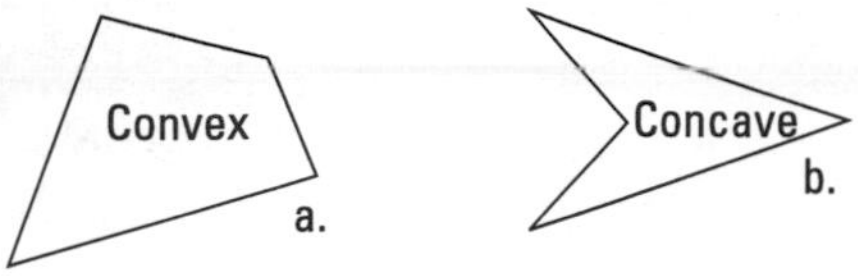

Figure 7-2:
Figure a
shows a
convex
quad, and
Figure b
shows a
concave
one.

Taking measure

Whatever your quad looks like, its interior angles have a sum measure and its exterior angles have a sum measure. Because a quad is a polygon, the sum of the exterior angles always equals 360° (the sum measure of the exterior angles of a polygon regardless of its number of sides always equals 360°). The sum of the interior angles is a different story. To find the sum of the measures of the interior angles of a quad, you need to use a formula that applies to polygons: $180(n-2)$, where n is the number of sides of the polygon. For a quadrilateral, n is always 4 (see Figure 7-3).

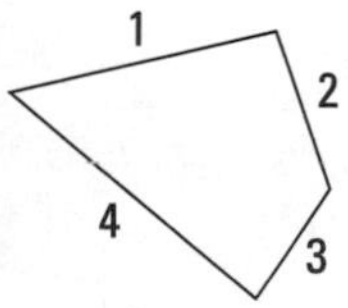

Figure 7-3.
To
determine
the sum
measure of
the interior
angles of a
quadri-
lateral, use
the formula
$180(n-2)$,
where n is
the number
of sides.

If n always equals 4, then $180(4-2)$ equals $180(2)$; 180 multiplied by 2 equals 360°. Exactly two diagonals can be drawn inside a quad. In Figure 7-4a, I have already drawn one. To draw the other diagonal, make a line that connects the other two vertices (see Figure 7-4b). The total number of diagonals in a polygon is determined by the formula $n(n-3)/2$. Remember, n is 4 (for the four sides of the quad), so $4(4-3)/2$. It follows that $4(1)/2$ or $4/2$. A quad has two diagonals. See Figure 7-4b.

A side note: The source of 360° becomes clear when you consider adding all the diagonals that can be drawn from a single vertex of a polygon. In the case of a quad, there is one diagonal that can be drawn from a single vertex. Now look at the diagonal in Figure 7-4a. Notice that it forms triangles inside the quad. Two triangles, to be exact. "So what?" you say. Ah, but nothing can be more exciting than this. You may know that the sum measure of the interior angles of a triangle is 180° (see Chapter 6). The quad has *two* triangles inside itself, so 180°(2) = 360°.

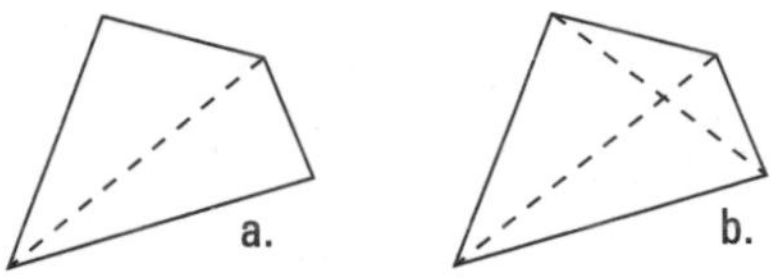

Figure 7-4:
In a quad, the total number of diagonals from a single vertex is one (Figure a). The total number of diagonals is two (Figure b).

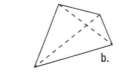

Parallelograms Contain Interesting Messages

Via special delivery . . . The #1 rule about a parallelogram, aside from the fact that it's a quad, is that it has opposite parallel sides (see Figure 7-5). And here's another message: Opposite sides of a parallelogram are equal. Yep, that's two pairs of congruent sides. And just when you think it can't get any better; consider this: There are two pairs of congruent opposite angles. Put all *that* in the blender, and out pours a quad with parallel congruent opposite sides and congruent opposite angles.

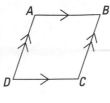

A parallelogram has two pairs of congruent angles that make up its four vertices. You name the parallelogram from these four vertices. Start at one vertex and use each consecutive vertex until none of them is left. The parallelogram in Figure 7-6, for example, is identified as □*ABCD*. Read this as "parallelogram *ABCD*."

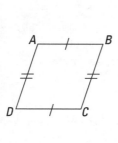

Theorem 7-1: If both pairs of opposite sides of a quadrilateral are congruent, the quadrilateral is a parallelogram.

Translation: If the opposite sides of a four-sided figure are congruent, then the quad is a parallelogram.

In Figure 7-6, $\overline{AB}$ is congruent to $\overline{CD}$ ($\overline{AB} \cong \overline{CD}$) and $\overline{BC}$ is congruent to $\overline{DA}$ ($\overline{BC} \cong \overline{DA}$).

Proof 7-1 shows that opposite sides of a parallelogram are congruent.

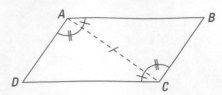

Given: $\overline{AB} \parallel \overline{DC}$ and $\overline{AD} \parallel \overline{BC}$

Prove: $\overline{AB} \cong \overline{DC}$ and $\overline{AD} \cong \overline{BC}$

Before you begin: Draw a line from vertex *A* to vertex *C*.

Proof 7-1:
The
Opposite
Sides of a
Parallel-
ogram
Proof

Statements	Reasons
1) $\overline{AB} \parallel \overline{DC}$ and $\overline{AD} \parallel \overline{BC}$	1) Given.
2) $\angle ACD \cong \angle BAC$ and $\angle DAC \cong \angle ACB$	2) Alternate interior angles of two parallel lines are congruent (see Chapter 2).
3) $\overline{AC} = \overline{AC}$	3) Reflexive (see Chapter 3).
4) $\triangle ADC \cong \triangle ABC$	4) ASA (see Chapter 6).
5) $\overline{AB} \cong \overline{DC}$ and $\overline{AD} \cong \overline{BC}$	5) CPCTC (see Chapter 6).

Theorem 7-2: If both pairs of opposite angles of a quadrilateral are congruent, the quadrilateral is a parallelogram.

Translation: If the opposite angles (those nonconsecutive angles) of a four-sided figure are congruent, then the quad is a parallelogram.

In Figure 7-6, angle *A* is congruent to $\angle C$ ($\angle A \cong \angle C$) and angle *B* is congruent to $\angle D$ ($\angle B \cong \angle D$).

Proof 7-2 shows that opposite angles of a parallelogram are congruent.

Diagonals with a twist

When diagonals are added to a polygon, you get more to work with. The parallelogram is no exception. When nonconsecutive vertices are joined with a diagonal, each of the two diagonals separates the parallelogram into two congruent triangles. At the point where the diagonals intersect, each diagonal is bisected (see Figure 7-7).

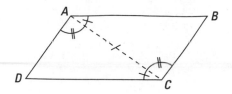

Given: $\overline{AB} \parallel \overline{DC}$ and $\overline{AD} \parallel \overline{BC}$

Prove: $\angle D \cong \angle B$ and $\angle A \cong \angle C$

Before you begin: Draw a line from vertex A to vertex C.

Proof 7-2:
The
Opposite
Angles of a
Parallel-
ogram
Proof

Statements	Reasons
1) $\overline{AB} \parallel \overline{DC}$ and $\overline{AD} \parallel \overline{BC}$	1) Given.
2) $\angle ACD \cong \angle BAC$ and $\angle DAC \cong \angle ACB$	2) Alternate interior angles of two parallel lines are congruent (see Chapter 2).
3) $\angle DAC + \angle BAC \cong \angle ACD + \angle ACB$	3) Angle Addition.
4) $\angle DAB \cong \angle DCB$ or $\angle A \cong \angle C$	4) Substitution.
5) $\overline{AC} = \overline{AC}$	5) Reflexive (see Chapter 3).
6) $\triangle ADC \cong \triangle ABC$	6) ASA (see Chapter 6).
7) $\angle D \cong \angle B$	7) CPCTC (see Chapter 6).

Figure 7-7:
The
diagonal
of a parallel-
ogram splits
the parallel-
ogram
into two
congruent
triangles.

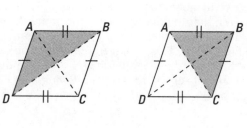

Theorem 7-3: A diagonal of a parallelogram divides the parallelogram into two congruent triangles.

Translation: When a single diagonal is drawn to opposite angles of a parallelogram, it creates two triangles of equal size.

In Figure 7-7, diagonal $\overline{AC}$ splits the parallelogram into two triangles — $\triangle ABC$ and $\triangle ADC$ — by the side-side-side (SSS) method (see Chapter 6).

Proof 7-3 shows that a diagonal splits a parallelogram into two congruent triangles.

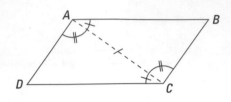

Given: $\overline{AB} \parallel \overline{DC}$ and $\overline{AD} \parallel \overline{BC}$

Prove: $\triangle ADC \cong \triangle ABC$

Before you begin: Draw a line from vertex A to vertex C.

Proof 7-3:
The Parallelogram Congruent Triangles Proof

Statements	Reasons
1) $\overline{AB} \parallel \overline{DC}$ and $\overline{AD} \parallel \overline{BC}$	1) Given.
2) $\angle ACD \cong \angle BAC$ and $\angle DAC \cong \angle ACB$	2) Alternate interior angles of two parallel lines are congruent (see Chapter 2).
3) $\overline{AC} = \overline{AC}$	3) Reflexive (see Chapter 3).
4) $\triangle ADC \cong \triangle ABC$	4) ASA (see Chapter 6).

Proof 7-4 shows that the two diagonals of a parallelogram bisect each other. That doesn't mean, however, that the diagonals are the same length. If a parallelogram has no right angles, its consecutive angles aren't equal, but they are supplementary (see Figure 7-8). This arrangement makes one angle acute and the other obtuse. In this case, the diagonal that is opposite the obtuse angle is longer than the diagonal across from the acute angle. In Figure 7-9, $\overline{QN}$ has a longer length than $\overline{MP}$.

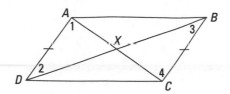

Given: $\overline{AB} \parallel \overline{DC}$ and $\overline{AD} \parallel \overline{BC}$

Prove: $\overline{AX} \cong \overline{CX}$ and $\overline{DX} \cong \overline{BX}$

Before you begin: Draw a line from vertex A to vertex C and from vertex D to vertex B.

Proof 7-4:
The Parallelogram Bisecting Diagonals Proof

Statements	Reasons
1) $\overline{AB} \parallel \overline{DC}$ and $\overline{AD} \parallel \overline{BC}$	1) Given.
2) $\angle 1 \cong \angle 4$ and $\angle 2 \cong \angle 3$	2) Alternate interior angles of two parallel lines are congruent (see Chapter 2).
3) $\overline{AD} \cong \overline{BC}$	3) Parallel lines are equidistant from each other (see Chapter 2).
4) $\triangle AXD \cong \triangle CXB$	4) ASA (see Chapter 6).
5) $\overline{AX} \cong \overline{CX}$ and $\overline{DX} \cong \overline{BX}$	5) CPCTC.

Figure 7-8:
In $\square MNPQ$, the sum of the measures of angles M and N equals 180°.

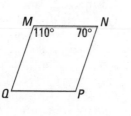

Figure 7-9:
If there's no
right angle,
then the
diagonal
opposite the
obtuse
angle is
longer than
the one
opposite the
acute angle.

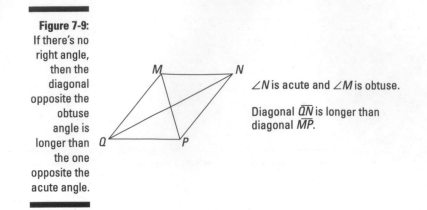

∠*N* is acute and ∠*M* is obtuse.

Diagonal $\overline{QN}$ is longer than
diagonal $\overline{MP}$.

Theorem 7-4: Consecutive angles of a parallelogram are supplementary.

Translation: When the measures of the two angles at each end of a side of a parallelogram are added together, they equal 180°.

In Figure 7-8, ∠*M* is supplementary to ∠*N* (∠*M* supp. ∠*N*) and ∠*Q* is supplementary to ∠*P* (∠*Q* supp. ∠*P*).

This area is double the space

Area is the amount of non-overlapping space in a polygon. When you draw a single diagonal in a parallelogram, you form two non-overlapping triangles. If the area of a triangle is ½*bh*, where *b* is the base and *h* is the height (altitude), then the area of a parallelogram is twice that. So (2)(½*bh*) equals *bh*. The area of a parallelogram is equal to the base multiplied by the altitude, plain and simple.

To calculate the area you first need to pick a base to work with. The base can be any side of a parallelogram. Next, you need to find the altitude that goes with the particular base you're working with. The *altitude* is a line that is (drawn) perpendicular to the base you've chosen and it extends to the opposite side. In Figure 7-10, the length of the base is 8 inches and the length of its corresponding altitude is 10 inches. The area is equal to (8)(10), or 80 square inches.

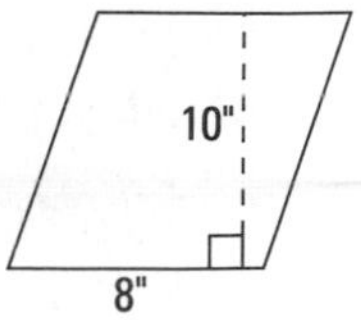

Figure 7-10:
Use $A = bh$ to find the area of a parallelogram.

10"

8"

Theorem 7-5: The area of parallelogram is equal to the product of the length of the base and the length of its corresponding altitude.

Translation: To get the area of a parallelogram, multiply the length of a side by the height of an altitude drawn to that side.

Parallelogram qualities in a nutshell

Want all the wonderful qualities of a parallelogram in one handy list? You got it:

- The opposite sides are parallel.
- The opposite sides are congruent.
- The opposite angles are congruent.
- The consecutive angles are supplementary.
- The diagonals bisect each other.
- Each diagonal creates two congruent triangles.
- All parallelograms are quadrilaterals.

Rectangles Aren't a Tangled Mess at All

A side skids into another side, and a chain reaction is set off. Sides and angles everywhere. Sounds like a mess, huh? But when you're referring to a rectangle, nothing could be further from the truth. Rectangles, affectionately known as *rects* by the geometry crowd, are neat and organized. They follow the rules of parallelograms. Being that a rectangle is a type of parallelogram, its opposite sides and angles are congruent, its opposite sides are parallel, and, most importantly, its consecutive angles are supplementary. To boot, a rectangle even has a few rules of its own.

What makes a rect special

What makes a rect special or different from the typical parallelogram? Is it the sides? Nope. It has to do with the angles. You need only show that one angle has a measure of 90°. Because consecutive angles are supplementary, everything falls into place from there. All four angles end up with measures of 90°. A rectangle has four right angles, which makes a rectangle equiangular (see Figure 7-11).

Figure 7-11:
A rectangle's trademark is its four right angles.

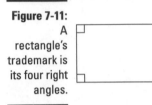

Theorem 7-6: All angles in a rectangle are right angles.

Translation: This one's pretty self-explanatory. A rectangle has four right angles. Each right angle is equal to 90°.

In Figure 7-11, each angle is a right angle and has a measure of 90°.

So what else is different? Well, because the angles of a rect are all congruent, and diagonals are drawn between opposite vertices, there must be some effect on the diagonals. And there is. They're also congruent. In Figure 7-12, the two diagonals are congruent to each other.

Figure 7-12:
The diagonals of a rectangle are congruent.

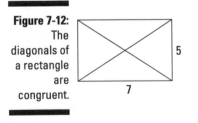

Theorem 7-7: The diagonals of a rectangle are congruent.

Translation: The diagonals of a rectangle are of equal length.

You can find the length of a diagonal of a rectangle by using the Pythagorean Theorem. Consider the diagonal as the hypotenuse of the triangle that's formed when a diagonal is drawn inside a rectangle. Use the equation $b^2 + h^2$ (where b is the length of the base and h is the length of the altitude) to find the length of the diagonal. Think of the length of the base as the *length* or b and the length of the altitude as the *width* or h. In Figure 7-12, $b = 7$ and $h = 5$. $7^2 + 5^2 = 49 + 25 - c^2 = 74$. The length of the diagonal equals $\sqrt{74}$ or 8.6 units.

The differences pointed out in Theorems 7-6 and 7-7 are key in proving that a rectangle is a parallelogram.

Proof 7-5 shows that the measure of each of the interior angles in a rectangle equals 90°, and Proof 7-6 shows that the two diagonals of a rectangle are of equal lengths.

Given: *WXYZ* is a rectangle.

Prove: $m\angle Z = 90°$

Proof 7-5: The Rectangle Interior Angles Proof

Statements	Reasons
1) A rectangle has four sides.	1) A rectangle is a quadrilateral.
2) *WXYZ* is a rectangle.	2) Given.
3) $180(n-2)$, where $n = 4$; $180(2) = 360°$	3) The formula for the sum of the measures of the interior angles of a four-sided polygon.
4) $360/4 = 90°$	4) A rectangle has four congruent angles. Each angle equals 90°.
5) $m\angle Z = 90°$	5) Substitution.

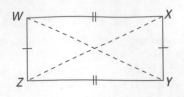

Given: *WXYZ* is a parallelogram. ∠*Z* is a right angle.

Prove: $\overline{WY} \cong \overline{ZX}$

Before you begin: Draw diagonals $\overline{WY}$ and $\overline{XZ}$.

Statements	Reasons
1) *WXYZ* is a parallelogram.	1) Given.
2) ∠*Z* is a right angle.	2) Given.
3) *WXYZ* is a rectangle.	3) A rectangle is a parallelogram one right angle.
4) ∠*W*, ∠*X*, ∠*Y*, ∠*Z* are right angles.	4) The four interior angles of a rectangle are right angles.
5) ∠*W* ≅ ∠*X* ≅ ∠*Y* ≅ ∠*Z*	5) All right angles are congruent.
6) $\overline{WZ} \cong \overline{XY}$	6) Opposite sides of a parallelogram are congruent.
7) $\overline{WX} \cong \overline{ZY}$	7) Same as #2.
8) $\overline{WZ} \cong \overline{WZ}$	8) Reflexive (see Chapter 3).
9) △*WXZ* ≅ △*WYZ*	9) SAS.
10) $\overline{WY} \cong \overline{ZX}$	10) CPCTC (see Chapter 6).

Proof 7-6:
The Rectangle Diagonals Proof

Rect space: Perimeter and area

A rect takes up space. The distance around the outside of the walls that contain this space is called the *perimeter*. The space inside the walls is the *area* and it is made up of square units. These square units are represented in Figure 7-13 as small boxes inside the perimeter and have equal heights and widths. The trick is to determine how many of these little boxes, or squares, exist within the perimeter of the polygon. If the walls of the polygon are perpendicular to each other, coming up with the answer is easy.

Figure 7-13:
The area of
a rect is
made up of
square units
that have
equal
heights and
widths.

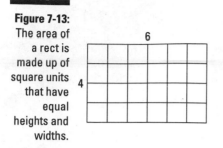

In Figure 7-13, six squares are across the top and four squares are down the side. The area of the rectangle is equal to the six squares multiplied by the four squares, or 24 square units.

Rect qualities in a nutshell

A rect isn't a wreck if it has the following qualities:

- The opposite sides are parallel.
- The opposite sides are congruent.
- The opposite angles are congruent.
- The consecutive angles are supplementary.
- The diagonals bisect each other.
- Each diagonal creates two congruent triangles.

The preceding six qualities are the same as those for a parallelogram. A rectangle adds its own flavor by having two more qualities, which follow.

- All four angles are right angles.
- The diagonals are congruent.

Rhombi Are Diamonds in the Rough

Have you ever played baseball? If you have, a rhombus (singular for *rhombi*) looks like a baseball diamond (see Figure 7-14). Like a rectangle, a rhombus is also a parallelogram — with a few additional stipulations. Its claim to fame is that it has two adjacent congruent sides.

Figure 7-14:
A baseball diamond is an example of a rhombus.

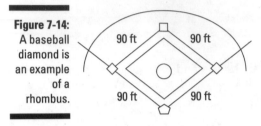

Because a parallelogram's opposite sides are congruent and a rhombus's adjacent side are congruent, it follows from here that all sides of a rhombus are congruent. This makes it an equilateral parallelogram. As you may know, diagonals are affected by changes in side length and number. The congruent sides of a rhombus allow the diagonals to bisect its (vertex) angles. The diagonals also pull double duty as perpendicular bisectors of each other.

Theorem 7-8: If a quadrilateral is equilateral, it is a rhombus.

Translation: A quad with four sides of equal length is a rhombus.

Proof 7-7 shows that a rhombus has equal sides.

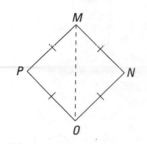

Given: $\overline{MP} \parallel \overline{NO}$, $\overline{MP} \cong \overline{MN}$, and $\overline{MO}$ bisects $\angle M$.

Prove: $\overline{MP} \cong \overline{NO}$

Before you begin: Draw a line from vertex M to vertex O.

Proof 7-7:
The
Rhombus
Equal Sides
Proof

Statements	Reasons
1) $\overline{MP} \parallel \overline{NO}$	1) Given.
2) $\overline{MN} \cong \overline{PO}$	2) Parallel lines are equidistant making $\overline{MN}$ and $\overline{PO}$ of equal length (see Chapter 2).
3) $\overline{MP} \cong \overline{MN}$	3) Given.
4) $\overline{MP} \cong \overline{PO}$	4) Substitution (see Chapter 3).
5) $\overline{MO} \cong \overline{MO}$	5) Reflexive (see Chapter 3).
6) $\overline{MO}$ bisects $\angle M$.	6) Given.
7) $\angle PMO \cong \angle OMN$	7) The definition of a bisector.
8) $\triangle OPM \cong \triangle MNO$	8) SAS (see Chapter 6)
9) $\overline{MP} \cong \overline{NO}$	9) CPCTC (see Chapter 6).

Theorem 7-9: The diagonals of a rhombus are perpendicular.

Translation: The diagonals of a rhombus meet to form right angles (angles with a measure of 90°).

Proof 7-8 shows that the diagonals of a rhombus are *orthogonal,* meaning that both are perpendicular to each other.

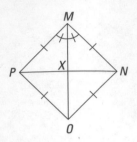

Given: *MNOP* is a rhombus, diagonals $\overline{MO}$ and $\overline{PN}$ intersect at *X*, and $\overline{MO}$ bisects $\angle M$.

Prove: $\overline{MO} \perp \overline{PN}$

Before you begin: Mark up the figure with the information given.

Proof 7-8:
The
Rhombus
Perpen-
dicular
Diagonals
Proof

Statements	Reasons
1) *MNOP* is a rhombus.	1) Given.
2) $\overline{PM} \cong \overline{ON} \cong \overline{PO} \cong \overline{MN}$	2) A rhombus is equilateral.
3) $\overline{MO}$ bisects $\angle M$.	3) Given.
4) $\angle PMX \cong \angle NMX$	4) The definition of a bisector.
5) $\overline{MX} \cong \overline{MX}$	5) Reflexive (see Chapter 3).
6) $\triangle MPX \cong \triangle MNX$	6) SAS (see Chapter 6).
7) $\angle MXP \cong \angle MXN$	7) CPCTC (see Chapter 6).
8) $\overline{MO} \perp \overline{PN}$	8) When two lines meet to form congruent adjacent angles, the lines are perpendicular.

Theorem 7-10: A parallelogram is a rhombus if the diagonals bisect the vertex angles.

Translation: A polygon with the properties of a parallelogram can be sub-grouped as a rhombus if the diagonals split the vertex angles into two equal angles.

Proof 7-9 shows that the diagonals of a rhombus bisect one pair of vertex angles.

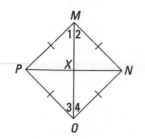

Given: *MNOP* is a rhombus.

Prove: $\angle 1 \cong \angle 2$ and $\angle 3 \cong \angle 4$

Before you begin: Mark up the figure with the information given.

	Statements	Reasons
Proof 7-9: The Rhombus Diagonals Bisect Vertex Angles Proof	1) *MNOP* is a rhombus.	1) Given.
	2) $\overline{PM} \cong \overline{ON} \cong \overline{OP} \cong \overline{MN}$	2) A rhombus is equilateral.
	3) $\angle P \cong \angle N$	3) Opposite angles of a rhombus are congruent.
	4) $\overline{MO} \cong \overline{MO}$	4) Reflexive (see Chapter 3).
	5) $\triangle OPM \cong \triangle ONM$	5) SSS (see Chapter 6).
	6) $\angle 1 \cong \angle 2$ and $\angle 3 \cong \angle 4$	6) CPCTC (see Chapter 6).

Theorem 7-11: The two diagonals of a rhombus form four congruent triangles.

Translation: The two diagonals of a rhombus intersect in a way that forms four congruent triangles. Take a look at Proof 7-10. This one is best explained with a visual.

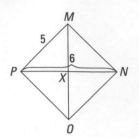

Given: *MNOP* is a rhombus, $\overline{PN} = 6$, and $\overline{PM} = 5$.

Prove: $\triangle PXM \cong \triangle MXN \cong \triangle NXO \cong \triangle OXP$

Before you begin: Label the lengths of the diagonals.

Proof 7-10:
The
Rhombus
Diagonals
Form
Congruent
Triangles
Proof

Statements	Reasons
1) *MNOP* is a rhombus.	1) Given.
2) $\overline{MN} = 5$, $\overline{NO} = 5$, $\overline{OP} = 5$, $\overline{PM} = 5$	2) A rhombus is equilateral.
3) $\overline{PN} = 6$	3) Given.
4) $\overline{PX} = 3$ and $\overline{NX} = 3$	4) Diagonals in a rhombus bisect each other.
5) $\overline{PM} = 5$	5) Given.
6) $\overline{PN}$ and $\overline{MO}$ are $\perp$.	6) Diagonals of a rhombus are perpendicular.
7) $\angle MXP$, $\angle MXN$, $\angle NXO$, and $\angle PXO$ are right angles.	7) Perpendiculars form right angles.
8) $\triangle PXM$, $\triangle NXM$, $\triangle PXO$, and $\triangle NXO$, are right triangles.	8) A right triangle has a right angle.
9) $\triangle PXM \cong \triangle MXN$	9) Hypotenuse Leg.
10) $\triangle PXM \cong \triangle MXN \cong \triangle NXO \cong \triangle OXP$	10) Hypotenuse Leg.

How many carats (area stuff)

How big is that diamond? The area of a rhombus revolves around the measures of its diagonals. To find the area of a rhombus, multiply the lengths of the two diagonals together and multiply that product by one-half (see Figure 7-15). If you're more comfortable with division, you can divide the product of the diagonal lengths by 2. Either way, the result is the same.

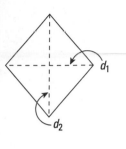

Figure 7-15:
The area of a rhombus is $\frac{1}{2}((d_1)(d_2))$, where d_1 is one diagonal and d_2 is the other.

Theorem 7-12: The area of a rhombus is one-half the product of the lengths of the two diagonals, or $A = \frac{1}{2}((d_1)(d_2))$.

Translation: Plug the numbers into the formula, and you get the area of a rhombus. Note that d_1 is one diagonal and d_2 is the other.

Rhombi qualities in a nutshell

The rhombus has all the qualities of a parallelogram plus some:

- ✔ The opposite sides are parallel.
- ✔ The opposite sides are congruent. (Actually, in the case of the rhombus, all four sides are congruent.)
- ✔ The opposite angles are congruent.
- ✔ The consecutive angles are supplementary.
- ✔ The diagonals bisect each other.
- ✔ Each diagonal creates two congruent triangles.

The preceding six qualities are the same as those for a parallelogram. A rhombus adds its own flair for originality by adding two more qualities, which follow.

- ✔ The diagonals are perpendicular to each other (they're *orthogonal*).
- ✔ The diagonals bisect the vertex angles.

It's Hip to Be Square

Here's a riddle for you: What has four sides; is a parallelogram because, among other things, its opposite sides are parallel; has the right angles of a rectangle; and has the four congruent sides of a rhombus? Take a peek at Figure 7-16 to find out.

Figure 7-16:
A square.

Yep, that figure's showing a square. To show that a quad is a square, you have a choice: You can show that the quad is a rectangle that has congruent adjacent sides, or you can show that it's a rhombus with a right angle. Which path you choose depends on what information you have about the polygon.

Theorem 7-13: A square is an equilateral quadrilateral.

Translation: A square has four equal sides.

Proof 7-11 shows that a square has four equal sides.

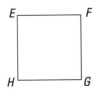

Given: *EFGH* is a rectangle and $\overline{EH} \cong \overline{EF}$.
Prove: $\overline{EH} \cong \overline{EF} \cong \overline{FG} \cong \overline{HG}$

Statements	Reasons
1) *EFGH* is a rectangle.	1) Given.
2) $\overline{EH} \cong \overline{EF}$	2) Given.
3) $\overline{EH} \cong \overline{FG}$	3) Opposite sides of a rectangle are congruent.
4) $\overline{EF} \cong \overline{HG}$	4) Same as #3.
5) $\overline{EH} \cong \overline{EF} \cong \overline{FG} \cong \overline{HG}$	5) Substitution.

Proof 7-11:
The Square
Four Equal
Sides Proof

Theorem 7-14: A square is a rhombus with one right angle.

Translation: A square is a rhombus with one right angle. So, all its angles are right angles because opposite angles are equal. A rhombus has equal sides. Put all this info together, and you get a square.

Proof 7-12 shows that if a rhombus has one right angle, then it's a square.

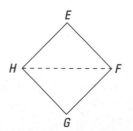

Given: *EFGH* is a rhombus, $\overline{HF}$ is a diagonal, and m∠*EHF* = 45°.

Prove: ∠*E* is a right angle.

Proof 7-12:
The Rhombus With a Right Angle is a Square Proof

Statements	Reasons
1) *EFGH* is a rectangle.	1) Given.
2) $\overline{HF}$ is a diagonal.	2) Given.
3) $\overline{HF} \cong \overline{HF}$	3) Reflexive.
4) m∠*EHF* = 45°	4) Given.
5) $\overline{EH} \cong \overline{EF}$	5) A rhombus is equilateral.
6) △*HEF* is an isosceles traingle.	6) An isosceles triangle has two congruent sides.
7) ∠*EHF* ≅ ∠*EFH*	7) Base angles opposite congruent sides of a triangle are congruent.
8) m∠*EFH* = 45°	8) Substitution.
9) m∠*EFH* + m∠*EHF* = 90°	9) Addition.
10) m∠*E* = 90°	10) The sum measure of the interior angles of a triangle is 180°.

Using the diagonals

Diagonals give you additional information. All you need is the length of one side of a square, and you can get the length of the diagonal. Just use the formula $D = s\sqrt{2}$, where D is the length of the diagonal and s is the length of any side (see Figure 7-17). Pick a side, it doesn't matter which one. It's a square. All sides have the same length. In Figure 7-17, if s equals 10, then the measure of the diagonal is $10\sqrt{2}$, or approximately 14.14 units.

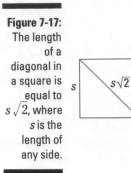

Theorem 7-15: The measure of a diagonal in a square is the length of any side multiplied by the square root of 2.

Translation: The length of a diagonal in a square equals $s\sqrt{2}$, where s is the length of any side.

Squaring off the perimeter and cordoning off the area

The perimeter of a square is equal to the sum of the lengths of all four sides, or $s + s + s + s = P$ (where s is the length of each side). To save space, you can also write the formula as $4s$ (see Figure 7-18). So, if s equals 8, for example, then the length of the perimeter equals 32 units.

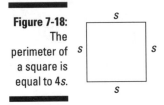

You can obtain the area of a square in two ways. Which way you choose depends on the information you have. One way deals with the measure of a side, and the other way deals with the measure of a diagonal.

The area of a square is actually its height times its width. To find the area of a square, you can multiply the measure of one side by the measure of another side. But because you're working with a square and both numbers are the same, the formula can be written as $A = s^2$. In Figure 7-19, the measure of a side equals 7. When you plug this number into the formula, the area of this square is 7^2 or 49 square units.

Figure 7-19:
Using the length of a side, you can use the formula $A = s^2$ to get the area of a square.

7

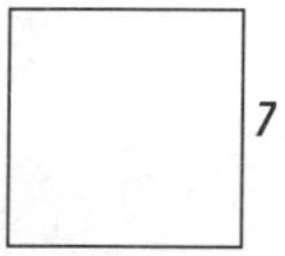

If you only have the measure of the diagonal, have no fear. You can still get the area by multiplying the square of the length of the diagonal by one-half — $A = \frac{1}{2}d^2$. In Figure 7-20, the length of the diagonal is 6. When you plug this number into the formula, you find that the area of this square is ½(36), or 18 square units.

Figure 7-20:
Using the length of a diagonal, you can use the formula $A = \frac{1}{2}d^2$ to get the area of a square.

$d = 6$

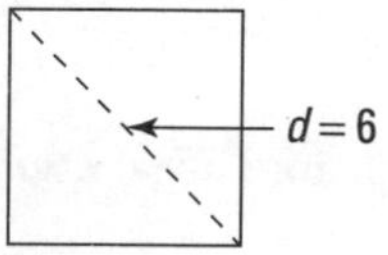

Square qualities in a nutshell

A square suffers from multiple personality disorder. A square is first a quadrilateral, but it is also a parallelogram, a rectangle, and a rhombus. A square is a regular polygon and exhibits characteristics of equilateral-ness and equiangular-ness. All its sides are equal, and each meets its adjacent side at a right angle.

Trapezoids Have Escape Hatches

Escaping from this shape may be a little more difficult than you think. The only given about a trapezoid is that it has one pair of parallel sides — one and only one pair. The parallel sides are called *bases,* and the nonparallel sides are *legs* (see Figure 7-21).

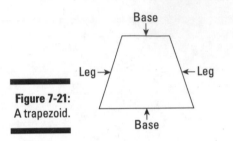

Figure 7-21:
A trapezoid.

A trapezoid that has congruent base angles is special. It's an isosceles trapezoid. And just like an isosceles triangle, an isosceles trapezoid has two congruent sides, not to mention that its opposite angles are supplementary (see Figure 7-22).

Figure 7-22:
An
isosceles
trapezoid.

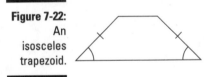

Theorem 7-16: The base angles of an isosceles trapezoid are congruent.

Translation: The base angles of an isosceles trapezoid have the same degree measures.

Proof 7-13 shows that the base angles of an isosceles trapezoid are congruent.

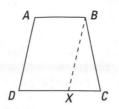

Given: *ABCD* is a trapezoid, $\overline{AB} \parallel \overline{DC}$ and $\overline{AD} \cong \overline{BC}$.

Prove: $\angle A \cong \angle ABC$

Before you begin: Draw a line that is parallel to $\overline{AD}$ and name it $\overline{BX}$.

Statements	Reasons
1) *ABCD* is a trapezoid.	1) Given.
2) $\overline{AB} \parallel \overline{DC}$ and $\overline{AD} \cong \overline{BC}$	2) Given.
3) $\overline{AD} \parallel \overline{BX}$	3) Drawn to be parallel.
4) *ABXD* is a parallelogram.	4) Opposite sides of a parallelogram are parallel.
5) $\overline{AD} \cong \overline{BX}$	5) Opposite sides of a parallelogram are congruent.
6) $\overline{BC} \cong \overline{BX}$	6) Transitive (see Chapter 3).
7) $\angle BXC \cong \angle C$	7) Angles opposite congruent sides are congruent. (see Chapter 6).
8) $\angle D \cong \angle BXC$	8) If two parallel lines are cut by a transversal ($\overline{DC}$), corresponding angles are congruent (see Chapter 2).
9) $\angle D \cong \angle C$	9) Transitive.
10) $\angle D$ is supplementary to $\angle A$; $\angle C$ is supplementary to $\angle ABC$.	10) If two parallel lines are cut by a transversal, interior angles on the same side of a transversal are supplementary (see Chapter 2).
11) $\angle A \cong \angle ABC$	11) Supplements of congruent angles are congruent (see Chapter 2).

Proof 7-13:
The
Isosceles
Trapezoid
Congruent
Base
Angles
Proof

The median

The median of a trapezoid is parallel to each base (to the sides that are parallel to each other). The median is drawn from the midpoint of one nonparallel side to the midpoint of the other nonparallel side. To get the length of the median, add the lengths of the two bases and multiply that sum by one-half. The equation form is $\frac{1}{2}(b_1 + b_2)$. In Figure 7-23, one base has a length of 8, and the other base has a length of 10. When you plug these numbers into the formula, you get $\frac{1}{2}(8 + 10)$, or a length of 9 units.

Figure 7-23:
The length of the median of a trapezoid is $\frac{1}{2}(b_1 + b_2)$.

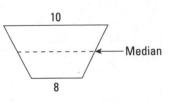

The diagonal

The diagonals of a run-of-the-mill trapezoid have nothing to note. The diagonals of an isosceles trapezoid have a short story: The diagonals are congruent.

Theorem 7-17: The diagonals of an isosceles trapezoid are congruent.

Translation: The diagonals of an isosceles trapezoid are the same length.

Proof 7-14 shows that the diagonals of an isosceles trapezoid are congruent.

The altitude

The altitude is the distance you have to climb to get up and out the top of the trapezoid. Consider the altitude to be a rope dropped from the upper base to the lower base. When the rope hits the lower base, it becomes perpendicular to it. You can draw many altitudes in a trapezoid, so there are many ways to escape.

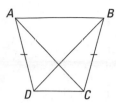

Given: *ABCD* is an isosceles trapezoid. $\overline{AC}$ and $\overline{DB}$ are diagonals.

Prove: $\overline{AC} \cong \overline{DB}$

Before you begin: Mark up the figure with the information given.

Statements	Reasons
1) *ABCD* is an isosceles trapezoid.	1) Given.
2) $\angle DAB \cong \angle ABC$	2) Angles opposite congruent sides are congruent.
3) $\overline{AC}$ and $\overline{DB}$ are diagonals.	3) Given.
4) $\overline{AB} \cong \overline{AB}$	4) Reflexive (see Chapter 3).
5) $\triangle ACD \cong \triangle BCD$	5) SAS (see Chapter 6).
6) $\overline{AC} \cong \overline{DB}$	6) CPCTC (see Chapter 6).

Proof 7-14: The Isosceles Trapezoid Congruent Diagonals Proof

The area

A diagonal divides a trapezoid into two triangles. These triangles aren't congruent, but they do represent two distinct spaces. You add together the areas on either side of the diagonal to define the whole area of the trapezoid. When the areas are put together, the entire trapezoid area is equal to the sum of the base lengths multiplied by the height of the trapezoid divided by two.

Theorem 7-18: The area of a trapezoid is equal to one-half the product of the length of the altitude multiplied by the sum of the bases.

Translation: $A = \frac{1}{2}h(b_1 + b_2)$.

In Figure 7-24, the bases are 8 units and 5 units, and the altitude is 6 units. The area of this trapezoid is equal to $\frac{1}{2}(6)(8 + 5)$, or 39 square units.

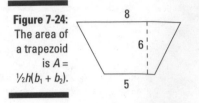

Figure 7-24: The area of a trapezoid is $A = \frac{1}{2}h(b_1 + b_2)$.

Go Fly a Kite

Yellow, pink, or blue swirling in the wind with a tail swishing behind . . . . Ah, memories of springtime. But this kind of kite is actually a type of quad. It does get its name, though, from the flying amusement that you take out on a windy day (see Figure 7-25).

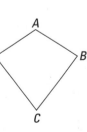

Figure 7-25: The kite actually gets its name from the toy that flies at the end of a string.

A kite is a quad with two pairs of adjacent congruent sides (see Figure 7-26), and the measure of any angle formed by the intersection of two sides is less than 180°. To form a kite, paste two isosceles triangles with different side lengths together at their bases.

Figure 7-26: The kite is a quad with two pairs of adjacent congruent sides.

In Figure 7-26, side $\overline{DA}$ is congruent to side $\overline{AB}$, and side $\overline{DC}$ is congruent to $\overline{BC}$.

You know how regular kites have brace sticks on them that form a cross? Well, on a kite quad, those braces are actually the diagonals of the kite. The braces form a 'T' and are perpendicular to each other. In Figure 7-27, for example, $\overline{AC}$ is perpendicular to $\overline{DB}$.

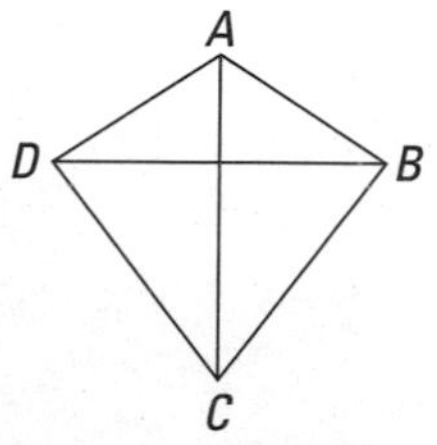

Figure 7-27:
The diagonals of a kite are perpendicular.

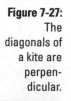

All Arrowheads Point This Way

Like the kite, the arrowhead quad gets its name from a real-world object. The arrowhead looks like an artifact found from prehistoric days — like the arrowhead a caveman used for hunting. The arrowhead shape in geometry is not an exact copy, but bears a good resemblance (see Figure 7-28).

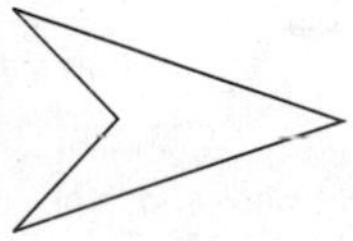

Figure 7-28:
The arrowhead quad looks like something cavemen used for hunting.

Geometrically, what's so special about an arrowhead? Consider the definition of a concave polygon: It has an angle of more than 180°. An arrowhead has one angle with a measure of more than 180°, so it's a concave quad. That kind of puts a dent in things.

Like any other quad, an arrowhead has diagonals. As usual, the diagonals are lines drawn to the opposite angles. The difference is that one diagonal of the arrowhead actually travels outside the polygon to connect the vertices. The diagonals of an arrowhead are perpendicular at the point of intersection, which may mean that the diagonals have to be extended outside the figure in order for them to intersect. That's the case with the diagonals shown in Figure 7-29. Diagonals $\overline{AC}$ and $\overline{BZ}$ have been extended, and they're perpendicular at point Z.

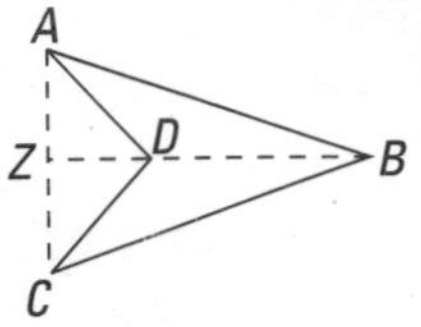

Figure 7-29: $\overline{AC}$ and $\overline{BZ}$ are the diagonals of this arrowhead.

Quad Aid

Figure 7-30 is your big-time visual aid. It shows how the various quads shape up relative to each other.

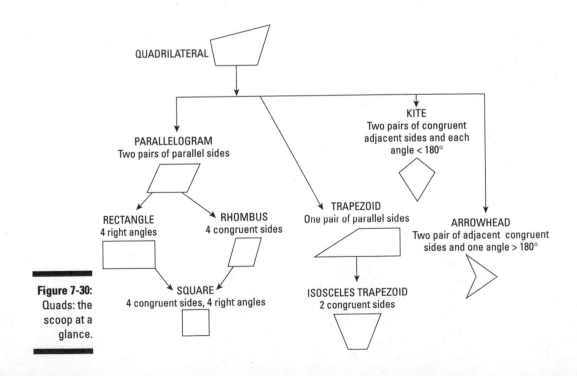

Figure 7-30: Quads: the scoop at a glance.

And if you just can't get enough about quads, Tables 7-1 and 7-2 give even more info about them. Table 7-1 compares quad properties. (***Note:*** It doesn't include trapezoids, kites, and arrowheads, though, because there isn't much conformist info for them.) Table 7-2 gives you the at-a-glance version of the formulae used with quads. (***Note:*** It's a Swiss cheese version. There are holes in it for the formulae that don't exist.)

Table 7-1:	Quad Properties at a Glance			
Properties	*Parallelogram*	*Rectangle*	*Rhombus*	*Square*
Has four sides	Yes	Yes	Yes	Yes
Opposite sides are parallel	Yes	Yes	Yes	Yes
Opposite sides are congruent	Yes	Yes	Yes (all four sides)	Yes (all four sides)
Opposite angles are congruent	Yes	Yes (all four)	Yes	Yes (all four)
Consecutive angles are supplementary	Yes	Yes	Yes	Yes
Has a right angle	No	Yes	No	Yes
Diagonals bisect each other	Yes	Yes	Yes	Yes
Diagonals bisect the vertex angles	No	No	Yes	Yes
Diagonals are congruent	No	Yes	No	Yes
Diagonals are perpendicular	No	No	Yes	Yes
Diagonals form two pairs of congruent triangles	Yes	Yes	Yes	Yes
Diagonals form four pairs of congruent triangles	No	No	Yes	Yes

Table 7-2:	Quad Formulae at a Glance		
Quadrilateral	*Area*	*Length of Diagonal*	*Length of median*
Parallelogram	$A = bh$		d or h
Rectangle	$A = bh$	$b^2 + h^2$	d or h
Rhombus	$A = \frac{1}{2}((d_1)(d_2))$		
Square	$A = s^2$	$D = s\sqrt{2}$	s
	$A = \frac{1}{2}d^2$		
Trapezoid	$A = \frac{1}{2}h(b_1 + b_2)$		$\frac{1}{2}(b_1 + b_2)$

Chapter 8

Going in Circles

You run into circles a lot in geometry. Ever seen a dog chase his tail? 'Round and 'round he goes. How's Rover ever supposed to get anywhere? Well, unless he looks up and heads off in another direction, he won't.

Unlike Rover, *you* can head off in a new direction in this chapter. You get the scoop on all kinds of circle stuff, like how to construct a circle using a compass and how to add lines, angles, and polygons to your masterpiece. *Voilà.* Then you'll be ready for an exhibition at the Metropolitan Museum of Art.

"Excuse Me, Can You Tell Me Why a Circle Is Round?"

Picture a circle in your head. I'm sure the first image that comes to mind is something round. Well, how did it get to be round? Next time someone asks, you'll have an answer: It involves the relationship between a point at the middle of the circle and the points that make up the circle. Figure 8-1 shows a point labeled with the capital letter M. Point *M* is the center of the circle. All points that make up the circle are of equal distance from this center point. To identify a circle, you use a symbol that looks like a bull's-eye. Because the center point is the center of attention, it can't go without mention. Hey, that rhymes. Cool. Anyway, the circle in Figure 8-1 is identified by the notation ⊙*M*.

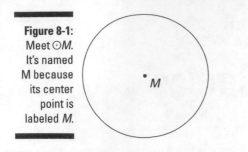

Now that you're armed with a definition for a circle, you can start construction. So put on your hardhat and break out your compass. To draw a circle, follow these steps:

1. **Adjust the compass so that the distance from the metal point to the pencil tip is half the width of the circle you want to draw.**

2. **Place the metal tip of the compass on the paper where you want the center of the circle to be.**

   Placing the metal point on the paper creates a tiny impression in the paper. This is the center point of the circle. Just don't push too hard, or you may rip a big hole in the paper. I've done that a few times. It's not so bad if it's the first circle on the page, but after you've drawn a few, you're really not going to want to start over. Guess there's always tape.

3. **With the tip of the pencil in contact with the paper, twirl the compass all the way around one time until the line connects to form a closed figure.**

   That's all there is to it (see Figure 8-2).

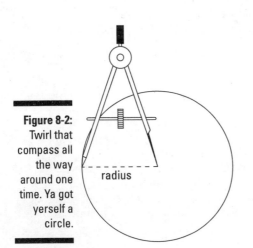

radius

Highways, Schmyways — Use the Thruways of a Circle

The thruways of a circle create its transportation network. They provide a way of getting from one point to another. These roads or paths aren't used in the construction of the circle itself but are built within, through, or just touching it. Thruways are grouped by where they let the traveler voyage in relation to the circle. In this section, I describe the various arteries of circle voyages, including radii, chords, diameters, secants, and tangents.

Radii: Voyage from the center of the circle

Cut to Rover in his circular yard. If Rover runs from the center of the yard-circle to any point at the fence, he has traveled along the *radius* of that circle. All similar journeys within a given circle starting at the center point and having a destination of the circle's edge are of the same distance. In Figure 8-3, Rover could run from point M to point Y, or he could run from point M to point X. Either way, he has traveled the same distance because $\overline{MY}$ and $\overline{MX}$ are both radii (the plural of radius) of circle M, and all radii of a circle are congruent.

Figure 8-3: $\overline{MY}$ and $\overline{MX}$ are both radii of the circle.

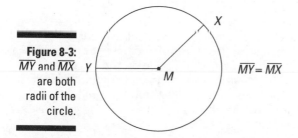

$\overline{MY} = \overline{MX}$

Theorem 8-1: All radii of a given circle or congruent circles are congruent.

Translation: All radii of the same circle are equal to each other. Also, if a circle is equal in size to another circle, then all the radii of the circles are equal to each other.

If two different circles have radii that are congruent to each other, then the circles are also congruent to each other. That's like having two backyards the same size. So if Rover goes for his run in either yard, his run is the same distance regardless of whether he's home or in the neighbor's yard. The flip-flop is also true; two circles are considered congruent if and only if their radii are congruent (see Figure 8-4). If Rover's run is the same distance in both backyards, then the backyards are the same size.

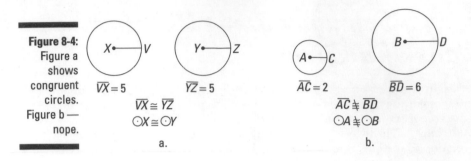

Figure 8-4:
Figure a
shows
congruent
circles.
Figure b —
nope.

$\overline{VX} = 5$ $\overline{YZ} = 5$ $\overline{AC} = 2$ $\overline{BD} = 6$
$\overline{VX} \cong \overline{YZ}$ $\overline{AC} \not\cong \overline{BD}$
$\odot X \cong \odot Y$ $\odot A \not\cong \odot B$
a.   b.

Postulate 8-1: Two circles are considered congruent if and only if their radii are congruent.

Translation: Two circles are considered to be the same size if the radii of the circles are equal to each other.

In or out? By knowing the length of the radius of a circle, you can tell if a point is inside or outside a circle. If a point is located at a distance from the center less than the radius, it's inside. A point that lies at a distance from the center greater than the radius is outside.

Chords: In both major and minor keys

Rover takes his midmorning run. He runs along a path that extends from point *A* to point *B*. This path is known in the world of geometry as a *chord.* A chord differs from a radius in that both its endpoints lie on the circle. A chord divides the circle into two segments — a minor segment and a major segment. And they're just like what you're thinking: A minor segment is the smaller section of the circle cut off by the chord, and the major segment is the bigger section (see Figure 8-5).

Figure 8-5:
A minor
segment is
the smaller
portion of a
circle that's
cut off by a
chord.

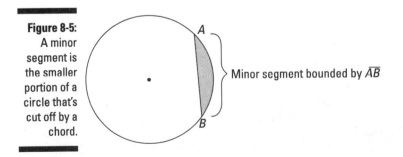

Minor segment bounded by $\overline{AB}$

Postulate 8-2: In the same circle or congruent circles, chords of equal lengths cut off equal arcs.

Translation: In a circle or equal circles, if two chords are of the same length, then the arcs (or segments on the circle) that each chord cuts off on the circle are equal.

Theorem 8-2: In the same circle or congruent circles, chords that are equidistant from the center of the circle are equal. The converse is also true: Equal chords are equidistant from the center of the circle.

Translation: In a circle or equal circles, chords that are the same distance from the center of the circle are equal to each other.

I'm going to hold off on including a proof for Theorem 8-2. Having a bit more info on circles will make understanding a proof for this theorem easier. If you can't wait, take a peek at Proof 8-10 . . . or you can wait until you get there.

Diameters: Equal pieces ye shall make

It's time for his midday run. Rover runs in a straight line from point Y to point Z (see Figure 8-6), and he runs right through the center of the circle on his way to the other side. Because his path extends from one side of the circle to the other side and passes through the center point, he has run across the *diameter*. The diameter slices the circle into exactly two equal pieces. So, if you have one cookie, and you break it along the diameter, both you and your friend get equal-sized pieces.

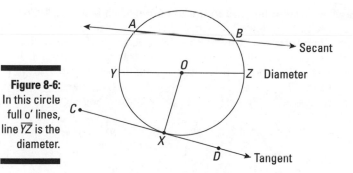

Figure 8-6:
In this circle
full o' lines,
line $\overline{YZ}$ is the
diameter.

The diameter is the longest path, or chord, across a circle and is equal to a distance twice that of the radius. The formula for calculating the diameter is $D = 2r$, where D stands for diameter and r stands for radius. In Figure 8-6, $\overline{YZ}$ is the diameter.

Theorem 8-3: The diameter is twice the distance of the radius.

Translation: Two radii equal one diameter because the radius only travels to the middle of the circle. That is, it travels half the distance across the circle, and two half distances make one whole distance.

Theorem 8-4: All diameters of a given circle are congruent.

Translation: All diameters in a circle are equal to each other.

Secants: Going in, going out again

It's midafternoon, and Rover has decided to stretch his legs. He enters the backyard, runs across it, and exits into the neighbor's yard to visit the neighbor's dog, Fido. Rover has run across a secant. A *secant* is a line (or segment or ray) that intersects the circle at two distinct points. Every secant line contains a chord of the circle. In Figure 8-6, $\overline{AB}$ is a chord on a secant line. If a secant contains the center point of the circle, then the chord portion of the secant is also the diameter.

Tangents: They always come a knockin' but never come in

Rover is being naughty — running around the neighborhood from backyard to backyard. Then he reaches the fence of his own backyard and decides to continue on without entering the yard. Rover has run a line tangent to his backyard. A *tangent* is a line (or segment or ray) that intersects (or touches) a circle in exactly one point. This point of contact is called the *point of tangency*. A tangent only touches on the circle; it never enters the circle. It knocks on the door but doesn't come in. In Figure 8-6, *X* is the point of tangency.

Postulate 8-3: Except for the point of tangency, all points on a tangent lie outside the circle.

Translation: A tangent doesn't enter into a circle. A tangent only touches the circle at one point. All other points on a tangent are outside the circle.

Theorem 8-5: A radius is perpendicular to a tangent at the point of tangency. That is, at the point of tangency, the measure of the angle that's formed by the intersection of a radius and a tangent is 90 degrees. In Figure 8-6, radius $\overline{OX}$ is perpendicular to $\overleftrightarrow{CD}$.

Translation: The radius and a tangent meet to form an angle of 90 degrees.

In Proof 8-1, I show you how to prove that a radius is perpendicular to a tangent at the point of tangency. I prove this one indirectly, meaning that I prove that certain statements are false in order to show that other statements must be true. I know, seems confusing. But by the time you complete the steps of the proof, you'll get the hang of it.

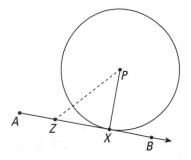

Given: $\overrightarrow{AB}$ is tangent to circle P at point X.

Prove: $\overline{PX} \perp \overrightarrow{AB}$ (This theorem is proved indirectly.)

Before you begin: Add a point Z to $\overrightarrow{AB}$. Connect point Z to point P.

Statements	Reasons
1) Either $\overline{PX} \perp \overrightarrow{AB}$ or $\overline{PX}$ not $\perp \overrightarrow{AB}$ (assume $\overline{PX}$ not $\perp \overrightarrow{AB}$	1) A statement is either true or false (see Chapter 3).
2) $\overline{PZ} < \overline{PX}$	2) If $\overline{PZ} \perp \overrightarrow{AB}$, $\overline{PZ}$ is the shortest distance from the center of the circle (point P).
3) $\overline{PZ}$ is not $< \overline{PX}$	3) Z is a point external to the circle P. $\overline{PX}$ is a radius with an endpoint on the circle. The assumption that $\overline{PX}$ is not $\perp \overrightarrow{AB}$ must be false.
4) $\overline{PX} \perp \overrightarrow{AB}$	4) $\overline{PX}$ is a radius with an endpoint on the circle and is the shortest distance between point P and point X. Therefore, $\overline{PX} \perp \overrightarrow{AB}$.

Proof 8-1:
The Perpendicular Radii Proof.

A circle has an infinite number of tangents, and the same line may be tangent to more than one circle. When two circles have the same tangent, the tangent line is called a common tangent.

A common tangent is either internal or external depending on how the tangent relates to the line of centers. Say what? The *line of centers* is the line that's drawn from circle center to circle center. Think of it like a line connecting the pupils of your eyes. If the common tangent crosses through the line of centers, then the tangent is an internal tangent (see Figure 8-7a). A common tangent that doesn't cross the line of centers is an external tangent (see Figure 8-7b). In Figure 8-7, $\overline{PX}$ is the line of centers.

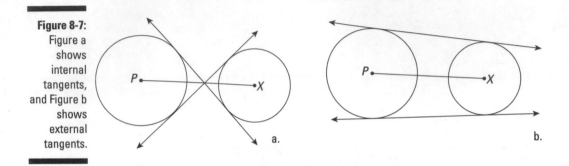

What's good for the lines is good for the circles. Just as lines can be grouped as internal or external, so can circles. It all starts with two circles that are tangent to each other (touch at one point) and a line that is tangent to both. If one circle is inside (or internal to) another so that their centers are on the same side of the tangent, then the circles are *internally tangent* (see Figure 8-8a). Circles whose centers lie on opposite sides of the tangent are *externally tangent* and look kind of like eyeglasses (see Figure 8-8b).

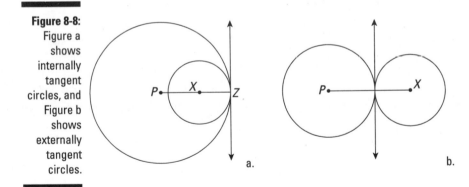

Theorem 8-6: If two circles are tangential to each other, then the line of centers connecting the two circles is perpendicular to their common tangent. In Figure 8-8a, $\overline{PZ}$ is perpendicular to the tangent line. In Figure 8-8b, $\overline{PX}$ is perpendicular to the tangent line.

Translation: If two circles touch each other at only one point, and a line is tangent to both circles at this same point, then a 90-degree angle is formed where the tangent and the line connecting the centers of the two circles meet.

In case you ever need to refresh your memory on circle lines stuff and really don't feel like wading through text, I've created Table 8-1, which is a summary of the line segment relationships in a circle. Ugh, what a mouthful. Gonna try that again: Table 8-1 contains info on the lines of a circle. It works well as a reference, especially if you're in a hurry to solve your proof. There. *Much* clearer.

Table 8-1	The Lines of a Circle	
Type of Angle	*Figure*	*Segment Relationships*
Central		$\overline{XA} = \overline{XB}$
Inscribed		In general, $\overline{PS} \neq \overline{PT}$
Chord-chord		$\overline{AZ} \times \overline{ZB} = \overline{CZ} \times \overline{ZD}$
Secant-secant		$\overline{AG} \times \overline{AB} = \overline{CG} \times \overline{CD}$
Radius-tangent		$GH \perp HM$
Tangent-tangent		$JK = JN$

Everybody's Got an Angle (Even a Circle)

A circle can contain multiple lines that form angles. These angles are collectively referred to as the *angles of a circle*. Intuitive, eh? Anyway, these angles are grouped by the location of an angle's vertex or by the type of lines used to form the angle.

Central and inscribed angles

Think of the vertex of an angle as a meeting place. Central and inscribed angles are identified by the location of this meeting place in relation to the circle. If the meeting place is at the center of the circle, then the angle is a central angle. A *central angle* is an angle with a vertex at the center point of the circle and radii as sides. An *inscribed angle* is formed by two chords that have a meeting place at a point on the circle. Check out Figure 8-9 for examples of both kinds of angles.

Figure 8-9:
∠*APC* is an inscribed angle, and ∠*DXB* is a central angle.

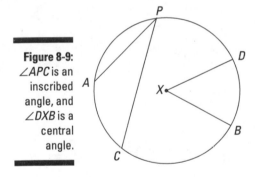

Angles of a circle can also be identified by their sides. You can easily identify these angle types by determining the types of lines that make up the sides of the angle. In addition to being formed by radii (central angles), angles can also be formed by chords, tangents, and secants. The vertex of these angles can be anywhere — on the circle, interior to the circle, or exterior to the circle.

Chord-chord angles

A *chord-chord angle* is just what it sounds like — an angle formed by the intersection of two chords in a circle (see the figure in Proof 8-3).

Theorem 8-7: If two chords intersect in a circle, the product of the length of the segments of one chord equals the product of the length of the segments of the other chord.

Translation: If two chords in a circle cross each other, if you multiply the lengths of the two segments of each line, the numbers will be equal.

In Proof 8-2, I show you how to prove that if two chords intersect in a circle, the product of the length of the segments of one chord equals the product of the length of the segments of the other chord.

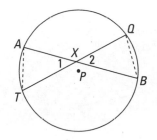

Given: Circle *P* with chords $\overline{AB}$ and $\overline{TQ}$ intersect at point *X*.

Prove $\overline{AX} \times \overline{XB} = \overline{TX} \times \overline{XQ}$

Before you begin: Draw *AT* and *QB* on the figure.

Proof 8-2:
The Chord
Segment
Proof

Statements	Reasons
1) $\angle 1 \cong \angle 2$	1) In $\triangle ATX$ and $\triangle QBX$, vertical angles are $\cong$ (see Chapter 3).
2) $\angle A \cong \angle Q$	2) Inscribed $\angle$s that intercept the same arc (*TB*) are $\cong$.
3) $\triangle TAX \cong \triangle ATX$	3) By the Angle-Angle Theorem of Similarity, if 2 $\angle$s of $\triangle \cong$ then $\triangle$s (see Chapter 10).
4) $\overline{AX}/\overline{XQ} = \overline{TX}/\overline{XB}$	4) Corresponding sides of similar $\triangle$s are proportional (see Chapter 10).
5) $\overline{AX} \times \overline{XB} = \overline{TX} \times \overline{XQ}$	5) Cross products.

Tangent-chord angles

When a tangent and a chord get together, a *tangent-chord angle* is formed (see Figure 8-10). The tangent can get together with a chord or the chord contained within a secant. The vertex of the angle is at the point of tangency.

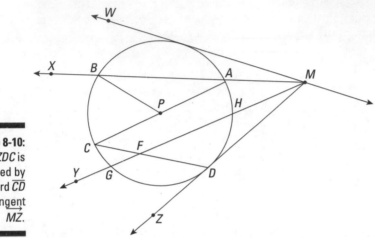

Figure 8-10: ∠*ZDC* is formed by chord $\overline{CD}$ and tangent $\overrightarrow{MZ}$.

Theorem 8-8: An angle formed by the intersection of a tangent and a chord is one-half the measure of the intercepted arc.

Translation: In Figure 8-10, m∠ZDC = ½m$\overparen{CD}$.

Secants and tangents (mix and match angles)

A secant-secant angle (formed by two secants), a tangent-secant angle (formed by a tangent and a secant), and a tangent-tangent angle (formed by two tangents) all meet outside the circle — or, more formally, have vertices that lie exterior to the circle. In Figure 8-10, for example, ∠*XMY* is formed by two secants, ∠*WMY* is formed by a tangent and a secant, and ∠*WMZ* is formed by two tangents.

Theorem 8-9: If two secants form an angle external to a circle, then the product of the length of one secant and the length of its internal segment is equal to the product of the length of the other secant and the length of its internal segment.

Translation: In Figure 8-10, $\angle XMY$ is formed by secants $\overrightarrow{XM}$ and $\overrightarrow{YM}$. So $\overline{BM} \times \overline{BA} = \overline{GM} \times \overline{GH}$.

In Proof 8-3, I show you how to prove that if two secants form an angle external to a circle, then the product of the length of one secant segment and its external segment is equal to the product of the length of the other secant segment and its external segment.

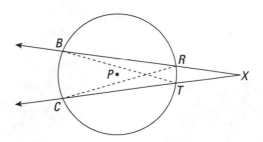

Given: $\overline{XB}$ and $\overline{XC}$ are secants to $\odot P$.

Prove: $\overline{XB} \times \overline{XR} = \overline{XC} \times \overline{XT}$

Before you begin: Draw $\overline{BT}$ and $\overline{CR}$ on the figure.

Proof 8-3:
The Secant
Segment
Proof

Statements	Reasons
1) $\angle XBT \cong \angle XCR$	1) $\angle$s that intercept the same arc are $\cong$.
2) $\angle X \cong \angle X$	2) Reflexive (see Chapter 3).
3) $\triangle XBT \sim \triangle XCR$	3) If two $\angle$s of a triangle $\cong$, the triangles are $\sim$ by Angle-Angle (AA) (see Chapter 6).
4) $\dfrac{\overline{XB}}{\overline{XC}} = \dfrac{\overline{XT}}{\overline{XR}}$	4) Corresponding sides of similar $\triangle$s are proportional (see Chapter 6).
5) $\overline{XB} \times \overline{XR} = \overline{XC} \times \overline{XT}$	5) Cross products are equal.

Theorem 8-10: If two tangents form an angle external to a circle, then the tangents are equal.

Translation: This one's pretty self-explanatory, but take a look at Proof 8-4 to run through the reasoning.

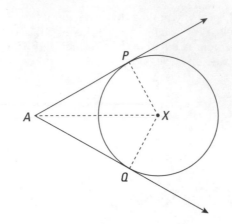

Given: $\overline{AP}$ and $\overline{AQ}$ are tangents to $\odot X$.

Prove: $\overline{AP} \cong \overline{AQ}$

Before you begin: Draw $\overline{XP}$, $\overline{XQ}$, and $\overline{XA}$ on the figure.

Proof 8-4:
The Equal
Tangents
Proof

Statements	Reasons
1) $\overline{XP} \cong \overline{XQ}$	1) All radii of a circle are congruent.
2) $\angle APX$ and $\angle AQX$ are right angles.	2) Tangents are perpendicular to a radius at the point of tangency.
3) $\overline{XA} \cong \overline{XA}$	3) Reflexive (see Chapter 3).
4) $\triangle AXP \cong \triangle AXQ$	4) Hypotenuse Leg (see Chapter 6; Postulate 6-6).
5) $\overline{AP} \cong \overline{AQ}$	5) Corresponding parts of congruent $\triangle$s are congruent (see Chapter 6).

Who Doesn't Love Pi?

If Rover were to run all the way around his circular backyard, he'd travel the linear distance around a circle. To determine this distance, you could open a circle at one place and stretch it to a straight line (see Figure 8-11).

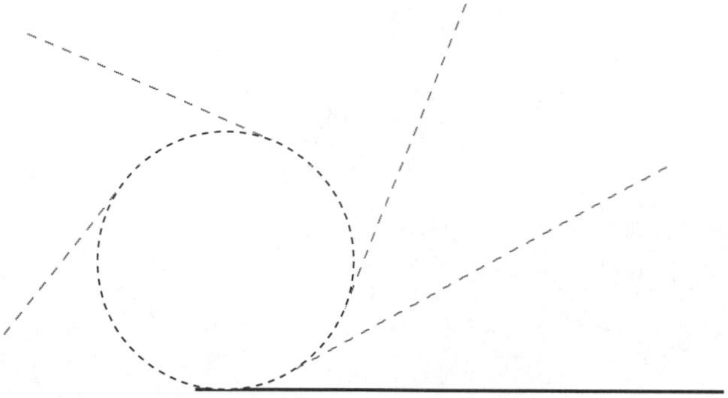

Figure 8-11:
St-r-r-r-ectching that circle to make a straight line, for measuring purposes.

But there's an easier way to get around! You can calculate this distance, known as the *circumference* of the circle, using an algebraic formula. But first, you need to know about a quantity that's used in the calculation. The quantity is known as *pi*. The symbol used to represent pi is the Greek letter p, which is written as π. The value of π is obtained by dividing the circumference of a circle by its diameter. π is a constant whose value is approximately equal to the decimal value of 3.14, or the improper fraction $\frac{22}{7}$. Once this value is substituted for π in a calculation, it's more appropriate to use $\approx$ in equations instead of = because 3.14 (or $\frac{22}{7}$) isn't the exact value of π but an approximation.

Enough of *that*. To calculate the circumference of a circle, use the formula $C = \pi D$ (where D is the diameter of the circle).

If only the radius distance is available, no need to fret. You can still use the formula $C = \pi D$. Remember that D (diameter) is equal to twice the distance of the radius, or $2r$. So it's not uncommon to see the formula for the circumference written as $C = 2\pi r$.

In Proof 8-5, I show you how to calculate the circumference of a circle. (This one is set up a little differently.)

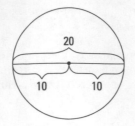

Before you begin: Develop a plan on how to solve the following equations using the information given.

Given: $D = 20$ (where D is the distance measure of the diameter):

$$C = \pi D$$

$$= \pi 20$$

$$= 20\pi$$

or

If D equals twice the distance as the radius, then if:

$$D = 20 \text{ then } r = D/2 = 20/2 = 10$$

$$C = 2\pi r$$

$$= 2\pi 10 = 2 \times 10 \times \pi$$

$$= 20\pi$$

Proof 8-5:
The Circle Circumference Proof

The Unsinkable Arc

No, not the boat. That ends with a *k*. This arc is the curved portion of a circle. If you think of a circle as a pie, the arc is the rounded crust around the edge. The *arc* is a set of points on the circle that are cut off, or intercepted, by an angle (see Figure 8-12). The sum of the measures of the consecutive arcs that form a circle is equal to exactly 360 degrees. So once around a circle is 360 degrees, or a whole pie is equal to 360 degrees.

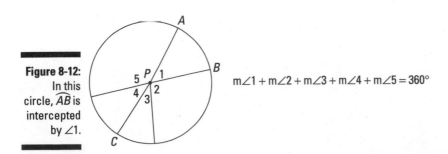

Figure 8-12:
In this circle, $\overarc{AB}$ is intercepted by $\angle 1$.

$$m\angle 1 + m\angle 2 + m\angle 3 + m\angle 4 + m\angle 5 = 360°$$

Theorem 8-11: The circumference of a circle is 360 degrees.

Translation: Once around a circle is 360 degrees.

For a central angle, you can easily determine the degree measure of an intercepted arc because the degree measure of the intercepted arc is equal to the degree measure of the central angle. No formulae involved — just a one-to-one relationship. In Figure 8-12, if the m∠APB equals 36 degrees, then the measure of $\overset{\frown}{AB}$ equals 36 degrees.

When the diameter (see $\overline{AC}$ in Figure 8-12) is the (straight) angle whose sides intercept (or cut off) the circle, the portion of the circle that is intercepted is referred to as a semicircle. The diameter cuts the circle into two equal halves. So, if a whole circle is 360 degrees, then the measure of a semicircle is 180 degrees. This line of thinking may help: Think of the diameter as being two radii. The angle formed by two radii along a diameter is a straight line. The measure of a straight line is 180 degrees.

Minor or major? Either way is no big deal

You can use a semicircle as an important reference point from which to categorize two types of arcs: major and minor. A *major arc* is an arc that's greater than a semicircle, or an arc that has a measure greater than 180 degrees. On the flip side, an arc that measures less than 180 degrees, or is less than a half circle, is a *minor arc*. Figure 8-13 shows both types of arcs.

Figure 8-13:
You've got two flavors or arcs to enjoy — major and minor.

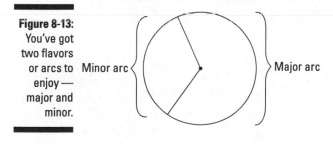

Minor arc   Major arc

Theorem 8-12: The length of an arc $= 2\pi r \times \left(\dfrac{n}{360 \ degrees} \right)$.

Translation: This is one of those formula things. Just plug in the information and you're set.

In Proof 8-6, I show how to obtain the length of an arc cut off by a 36° central angle. I'll do the math. This proof has a more informal structure since forcing it into the proof columns would just be too constricting and a bit stifling. It needs to breeeeeathe.

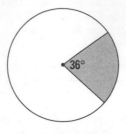

Before you begin: Develop a plan on how to solve the following equations using the information given.

Given: $r = 10$ (where r is the radius distance of the circle)
Degree measure of the arc = 36

Find: the length of arc cut off by a 36° central angle

$$\text{Length of an arc} = 2\pi r \times \left(\frac{n}{360}\right)$$

$$= 2\pi 10 \times (36/360)$$

$$= 20\pi \times (1/10)$$

$$= 2\pi$$

A proportion can also be used to achieve the same result:

$$\left(\frac{\text{Length of arc}}{\text{Circumference}}\right) = \left(\frac{\text{Degree measure of arc}}{360 \text{ degrees}}\right)$$

$$x/2\pi 10 = 36/360$$

$$x/20\pi = 1/10$$

$$10x = 20\pi \text{ (by cross products)}$$

$$10x/10 = 20\pi/10$$

$$x = 2\pi$$

Proof 8-6:
The Length
of Arc Proof

Finding the angle from the arc

The location of an angle's vertex relative to the circle determines how to obtain the angle's measure using its intercepted arc. Table 8-2 provides a quick summary for you to quickly determine how an angle is measured.

Table 8-2	Measuring an Angle		
Type of Angle	*Figure*	*Location of Vertex*	*Angle Measured By*
Central		At center (interior)	Intercepted arc $m\angle 1 = m\widehat{AB}$
Inscribed		On	One-half the intercepted arc $m\angle 2 = m\frac{1}{2}\widehat{ST}$
Inscribed chord-tangent			
Radius-tangent		On	Perpendicular $m\angle 3 = 90°$
Chord-chord		Interior	One-half the sum of intercepted arcs $m\angle 4 = \frac{1}{2}(m\widehat{AC} + m\widehat{DB})$
Secant-secant		Exterior	One-half the difference of the intercepted arcs $m\angle 5 = \frac{1}{2}(m\widehat{EF} - m\widehat{HK})$
Tangent-secant			
Tangent-tangent			

Theorem 8-13: The measure of an inscribed angle is one-half the measure of its intercepted arc.

Translation: Check out Table 8-2 for the formula and a figure.

In Proof 8-7, I show you how to prove that the measure of an inscribed angle is one-half the measure of its intercepted arc.

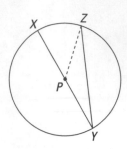

Given: $\angle XYZ$ is inscribed in $\odot P$.

Prove: $m\angle XYZ = \frac{1}{2}\, m\widehat{XZ}$

Before you begin: Draw $\overline{PZ}$ on the figure.

Statements	Reasons
1) $\overline{PY} \cong \overline{PZ}$	1) All radii of a circle are $\cong$.
2) $m\angle PZY = m\angle PYZ$	2) If two sides of a $\triangle$ are $\cong$, then $\angle$s opposite these sides are $\cong$ (see Chapter 6).
3) $m\angle PYZ = m\angle XYZ$	3) Reflexive.
4) $m\angle XPZ = m\angle XYZ + m\angle PZY$	4) The measure of an exterior $\angle$ of $\triangle$ = the sum of the measures of the two nonadjacent interior $\angle$ (see Chapter 6).
5) $m\angle XPZ = \frac{1}{2}m\angle XYZ + m\angle XYZ = 2m\angle XYZ$	5) Substitution (see Chapter 3).
6) $m\angle XPZ = \frac{1}{2}m\angle XYZ$	6) Proportions of equal $\angle$s are =. Therefore, halves of equals are =.
7) $m\angle XPZ = \widehat{XZ}$	7) The measure of a central $\angle$ = the measure of its intercepted arc.
8) $m\angle XYZ = \frac{1}{2}m\widehat{XZ}$	8) Substitution (see Chapter 3).

Proof 8-7:
The
Inscribed
Angle Proof

Theorem 8-14: The measure of an angle exterior to a circle that is formed by two secants is one-half the difference of its intercepted arcs.

Translation: Take a look at Table 8-2 for the formula and a figure. Also, check out Proof 8-8.

In Proof 8-8, I show you how to prove that the measure of an angle exterior to a circle that is formed by two secants is one-half the difference of its intercepted arcs.

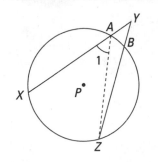

Given: $\overline{YX}$ and $\overline{YZ}$ are secants to $\odot P$.

Prove: $\angle Y = \frac{1}{2}\left(\widehat{XZ} - \widehat{AB}\right)$

Before you begin: Draw $\overline{AZ}$ on the figure.

Proof 8-8:
The Exterior
Angle
Measure
Proof

Statements	Reasons
1) $m\angle Y + m\angle Z = m\angle 1$	1) The measure of an exterior $\angle$ of a $\triangle$ = the sum of the measures of the nonadjacent interior $\angle$ (see Chapter 6).
2) $m\angle Y = m\angle 1 - m\angle Z$	2) Subtraction (see Chapter 3).
3) $\angle 1 = \frac{1}{2}\widehat{XZ},\ \angle Z = \frac{1}{2}\widehat{AB}$	3) An $\angle$ inscribed in a circle is measured by $\frac{1}{2}$ its intercepted arc.
4) $\angle Y = \frac{1}{2}\widehat{XZ} - \frac{1}{2}\widehat{AB}$ or $\angle Y = \left(\widehat{XZ} - \widehat{AB}\right)$	4) Substitution (see Chapter 3).

Congruent arcs

I have a round pizza pie. I cut the pizza into equal pieces from the center, and then the measures of the angles at the pointy tips of the pizza are the same for each piece. Because the measure of the arc is the same as the measure of the central angle, these slices have equal arcs as well. Arcs that are of equal measures are referred to as *congruent arcs*. Congruent arcs can be from the same circle or from congruent circles. So equal-size pizza slices can be from the same pizza or from two same-size pizzas.

Wow — was I hungry. I just ate all the pizza except for one slice. My sisters Sarah and Penny just came over, and they both want that last piece. So I do the only thing that's fair: Eat it myself. No, no. Just kidding. I find the middle of the crust and cut straight toward the pointy end of the slice. The crust is an arc, and by finding the middle, or midpoint, of this arc, I'm dividing it (and the pizza slice) into two congruent arcs. Doing so makes two equal-size pieces, so there's no arguing over who gets the bigger slice. If a single arc or slice of pizza is separated into two congruent arcs or slices, the cut made is called a *bisector* because it cuts through the arc at its midpoint (see Figure 8-14).

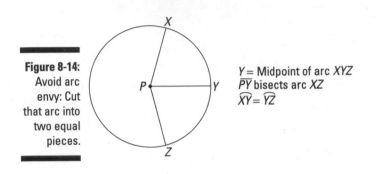

Figure 8-14:
Avoid arc
envy: Cut
that arc into
two equal
pieces.

Y = Midpoint of arc XYZ
$\overline{PY}$ bisects arc XZ
$\overset{\frown}{XY} = \overset{\frown}{YZ}$

Postulate 8-4: In the same circle or congruent circles, arcs of the same degree have the same length.

Translation: In the same circle or same-size circles, arcs with equal degree measures are also equal in length.

Theorem 8-15: If a diameter is perpendicular to a chord, then it bisects the chord and its arcs. The bisector creates congruent segments and congruent arcs.

Translation: If a diameter creates a 90-degree angle when it crosses a chord, the diameter cuts the chord into two equal segments and cuts its arc into two equal pieces as well.

In Proof 8-9, I show you how to prove that if a diameter is perpendicular to a chord, then it bisects the chord and its arcs, creating congruent segments and congruent arcs.

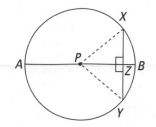

Given: Diameter: $\overline{AB}$ of $\odot P \perp \overline{XY}$

Prove: $\overline{XZ} \cong \overline{YZ}$, $\overparen{BX} \cong \overparen{BY}$, $\overparen{AX} \cong \overparen{AY}$

Before you begin: Draw $\overline{PX}$ and $\overline{PY}$ on Figure 8-24.

Proof 8-9:
The Congruent Segments and Arcs Proof

Statements	Reasons
1) $\overline{PX} \cong \overline{PY}$	1) All radii of a circle are $\cong$.
2) $\overline{AB} \perp \overline{XY}$	2) Given.
3) $\angle PZX$ and $\angle PZY$ are right angles.	3) Perpendiculars form right $\angle$.
4) $\overline{PZ} \cong \overline{PZ}$	4) Reflexive (see Chapter 3).
5) $\triangle PZX \cong \triangle PZY$	5) Hypotenuse Leg (see Chapter 6; Postulate 6-6).
6) $\overline{XZ} \cong \overline{YZ}$, $\angle XPZ \cong \angle YPZ$	6) Corresponding parts of congruent $\triangle$ are congruent (see Chapter 6).
7) $\overparen{XB} \cong \overparen{BY}$	7) Congruent central angles of a circle have congruent arcs.
8) $\overparen{AXB} \cong \overparen{AYB}$	8) A diameter bisects a circle.
9) $\overparen{AX} \cong \overparen{AY}$	9) Subtraction (see Chapter 3).

Theorem 8-16: In a circle, parallel chords cut off equal arcs.

Translation: Take a look at Proof 8-10 for the translation.

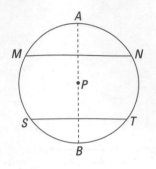

Given: In $\odot P$, $\overline{MN} \| \overline{ST}$

Prove: $m\overparen{MS} = m\overparen{NT}$

Before you begin: Draw diameter $\overline{AB} \perp \overline{MN}$ on the figure.

Proof 8-10:
The Parallel
Chords
Proof

Statements	Reasons
1) $\overline{MN} \| \overline{ST}$	1) Given.
2) $\overline{AB} \perp \overline{MN}$, therefore $\overline{AB} \perp \overline{ST}$	2) Substitution (see Chapter 3).
3) $m\overparen{MB} = m\overparen{NB}$	3) A diameter perpendicular to a chord bisects its arc.
4) $m\overparen{MS} = m\overparen{NT}$ $\quad m\overparen{MB} - m\overparen{SB} = m\overparen{MS}$ $\quad m\overparen{NB} - m\overparen{TB} = m\overparen{NT}$	4) Arc subtraction (see Chapter 3).

Finding the area of a sector of a circle

The *sector of the circle* is the region of the circle bounded by the radii that create the arc. You can visualize this one if you think of a circle as a whole pie and the sector as a slice of that pie. The bigger your slice of pie, the greater percentage of the whole you get to eat. You can figure out the area of your slice by setting up a proportion to determine the area of your slice relative to the area of the whole pie.

The degree measure of an arc is simply a percentage of the whole circle. This kind of ratio can also be applied to the area:

$\dfrac{Angle\ of\ sector}{Angle\ of\ circle} = \dfrac{Area\ of\ sector}{Area\ of\ circle}$ (where the angle of a circle equals 360, and the area (A) of a circle is equal to πr^2; and r is the radius of the circle).

Theorem 8-17: The area of a circle = πr^2

Translation: Yes, another formula. Take a look at Proof 8-12 for the translation.

In Proof 8-11, I show you how to calculate the area of a circle. It is a casual-style proof. The steps follow mathematically.

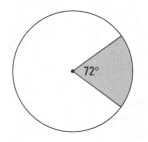

Given: $r = 10$ (where r is the length of the radius of a circle)
  $m\angle ABC = 72°$

Before you begin: Develop a plan on how to solve the following equations using the information given.

Find the area of the circle:

Formula for the Area of a Circle: $A = \pi r^2$

$$= \pi(10)^2$$

$$= \pi 100$$

$$= 100\pi \text{ square units}$$

Find the area of a sector of the circle:

Proportion for finding the area of a sector:

$$\left(\frac{\text{Angle of a sector}}{\text{Angle of a circle}}\right) = \left(\frac{\text{Area of a sector}}{\text{Area of a circle}}\right)$$

Angle of a sector $= 72°$ (given)

Angle of a circle $= 360°$

Area of the sector $= x$ (unknown)

Area of the circle $= 100\,\pi$

$$\frac{72}{360} = \frac{x}{100\pi}$$

$$7200\pi = 360x \text{ (cross multiply)}$$

$$\frac{7200\pi}{360} = \frac{360}{360}$$

$$20\pi = x$$

$$x = 20\pi \text{ square units}$$

Proof 8-11:
The Area of
a Circle
Proof

The Polygamous Circle: Relationships Galore

Circles are quite the social butterflies. They can have relationships to lines and angles. They can also have relationships to other polygons.

Inscribed circles: An insider's view

A circle can be placed inside a polygon. If the circle touches each side of the polygon at only one spot, the circle is *inscribed* within the polygon — meaning that because the circle and each side touch each other at one and only one point, each side of the polygon is tangential to the circle (see Figure 8-15).

Figure 8-15:
Inscribed circles are *inside* polygons.

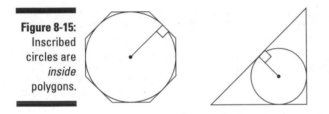

You know that there are always two sides to every story. The relationship between the circles and polygons shown in Figure 8-15 is no exception. The relationship can be described in two different ways depending on how you look at it: The circles are *inscribed* within the polygons, or the polygons *circumscribe* — completely surround — the circles. So, if a circle is inscribed within a given polygon, it's also true that the polygon surrounds the circle within its sides. As in life, which view you take depends on what information you have and are trying to prove.

Theorem 8-18: The sides of a circumscribed polygon are tangential to the inscribed circle.

Translation: An inscribed circle touches each side of the polygon in exactly one spot per side.

When tangential segments to a circle originate at a single point outside the circle, the lengths of the tangents from that external point to the point of contact with the circle are equal.

Take a look at Proof 8-12. $\odot O$ is inscribed in $\triangle ABC$. In this proof, you find the length of segments $\overline{CE}$, $\overline{BG}$, and $\overline{AF}$.

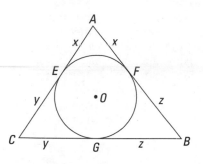

Given: $\odot O$ is inscribed in $\triangle ABC$.

$$\overline{AC} = 12, \overline{AB} = 13, \overline{BC} = 15$$

Find $\overline{CE}, \overline{BG}, \overline{AF}$

Before you begin: Notice the prove statement has been replaced with a "Find." Treat it the same. Use information from Theorem 8-10 "the lengths of the tangents from that external point to the point of contact with the circle are equal" to solve the proof.

Statements	Reasons
1) $\odot O$ is inscribed in $\triangle ABC$.	1) Given.
2) $\overline{AC} = 12, \overline{AB} = 13, \overline{BC} = 15$	2) Given.
3) $\overline{AE} = \overline{AF} = x$	3) If two lines tangent to a circle start at the same point outside the circle, then the lengths of the tangents from the external point to the point of contact with the circle are equal.
4) $\overline{CE} = \overline{CG} = y$	4) Same as #3.
5) $\overline{BG} = \overline{BF} = z$	5) Same as #3.
6) $z + y = 15$ $y + x = 12$ $\overline{z - x = \ 3}$	6) Subtract $\overline{AC}$ from $\overline{BC}$.
7) $z - x = 3$ $z + x = 13$ $2z = 16$ or $z = 8$	7) Addition (Add the equation from #6 to the measure of $\overline{AB}$).
8) $\overline{FB} = \overline{GB} = 8$	8) Substitution.
9) $\overline{GB} = \overline{BG}$	9) Identity.
10) $y = 15 - 8$ or $y = 7$ or $\overline{CE} = 7$	10) Substitution.
11) $x = 12 - 7$ or $x = 5$ or $\overline{AF} = 5$	11) Substitution.

Proof 8-12: The Equal Tangent Lengths from an External Point to a Circle Proof

Auxiliary lines are extra lines that can be added to a figure. And they're a geometric shape's best friend. They don't just go away because a shape is located inside or around another shape. In other words, inscribed and circumscribed shapes maintain their relationships to their auxiliary lines. Want proof? Take a look at the following theorem and proofs.

Theorem 8-19: The length of the radius of a circle inscribed in an equilateral triangle is one-third of the length of the altitude of the triangle.

Translation: If a circle is inscribed in an equilateral triangle, the radius of the circle is one-third the length of the height of the triangle.

In Proof 8-13, $\odot O$ is inscribed in $\triangle ABC$. Find the length of the radius given the measure of the side of an equilateral triangle. In Proof 8-14, $\odot O$ is inscribed in $\triangle ABC$. Find the length of the radius of the circle given the measure of the hypotenuse of a right triangle. In Proof 8-15, $\odot O$ is inscribed in $\triangle ABC$. Find the circumference of the circle given the lengths of the sides of the equilateral triangle in which the circle is inscribed.

The radius of an inscribed circle is equal to the length of the apothem of the circumscribed regular polygon (see Chapter 5 for a review of apothems). The area of a regular polygon is A = ½ap, where a is the length of its apothem and p is its perimeter. In Proof 8-16, $\odot Z$ is inscribed in polygon *ABCDEF*. Find the area of regular polygon *ABCDEF* given the radius of $\odot Z$. In Proof 8-17, $\odot O$ is inscribed in $\triangle ABC$. Find the difference in the areas between the circle and the triangle.

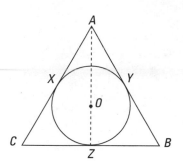

Given: ⊙ *O* is inscribed in equilateral △*ABC*.

The length of $\overline{AB}$ of △*ABC* is 5.

Find: $\overrightarrow{OZ}$ (the radius of ⊙ *O*)

Before you begin: Draw the altitude $\overline{AZ}$ to ∠A of △*ABC*.

<table>
<tr><td colspan="2">Statements</td><td colspan="2">Reasons</td></tr>
</table>

Proof 8-13: The Length of the Radius of an Inscribed Circle Proof

Statements	Reasons
1) ⊙ *O* is inscribed in equilateral △*ABC*	1) Given.
2) $\overline{AB} = 5$	2) Given.
3) $\overline{AZ}$ bisects vertex *A* and side $\overline{BC}$.	3) This is the altitude of △*ABC*.
4) $\overline{AZ} = 5\frac{\sqrt{3}}{2}$ or approximately 4.33	4) The altitude is $\frac{\sqrt{3}}{2}$ multiplied by the length of any side in an equilateral △ (see Chapter 6).
5) $\overline{OZ} = \frac{1}{3}(4.33)$ or approximately 1.44	5) The radius of a circle inscribed in an equilateral △ is ⅓ the length of the altitude of that △ .

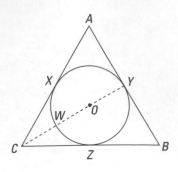

Given: ⊙ O is inscribed in equilateral △ABC.

The perimeter of △ACY is $14 + 7\sqrt{2}$. CW = 3.

Find: $\overline{YO}$ (the radius of ⊙ O)

Before you begin: Draw altitude $\overline{CY}$, which contains $\overline{YO}$, (the radius of ⊙ O).

Statements	Reasons
1) ⊙ O is inscribed in equilateral △ABC.	1) Given.
2) $\overline{AC} \cong \overline{BC} \cong \overline{AB}$	2) In an equilateral triangle, all sides are congruent (see Chapter 6).
3) The perimeter of △AYC is $14 + 7\sqrt{2}$.	3) Given.
4) △AYC is a right isosceles triangle.	4) Altitude $\overline{CY}$ forms two right isosceles triangles △AYC and △BYC within parent △ABC (see chapter 6).
5) For △AYC, $7 + 7 + 7\sqrt{2}$; where 7 equals the length of a side and $7\sqrt{2}$ is the length of the hypotenuse.	5) The measures of the sides of an isosceles right triangle follow formula $s + s + s\sqrt{2}$ (see Chapter 6).
6) $\overline{CY}$ is a median of $\overline{AB}$.	6) Line drawn as median (see Chapter 6).
7) $\overline{CY} = 7$	7) The length of the median to the hypotenuse of the right triangle is equal to ½ the length of the hypotenuse (see Chapter 6).
8) $\overline{CY} - \overline{CW} = 7 - 3 = 4$	8) The radius of a circle inscribed in an equilateral △ is ⅓ the length of its altitude (see Chapter 6).
9) $\overline{YO} = \frac{1}{2}(4)$ or 2.	9) Radius is ½ the diameter of a triangle (see Chapter 6).

Proof 8-14:
The Circle Inscribed in a Triangle Radius Length Proof

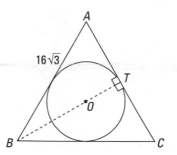

Given: $\odot O$ is inscribed in equilateral $\triangle ABC$.

$\overline{AB} = 16\sqrt{3}$

Find: The circumference of $\odot O$.

Before you begin: Draw $\overline{BT}$ as a perpendicular bisector to side $\overline{AC}$. Label the known segments.

Statements	Reasons
1) $\odot O$ is inscribed in equilateral $\triangle ABC$.	1) Given.
2) $\overline{AB} = 16\sqrt{3}$	2) Given.
3) $\overline{BC} \cong \overline{AB} \cong \overline{AC}$	3) All sides of an equilateral triangle are congruent (see Chapter 6).
4) $\overline{BT}$ bisects $\angle C$	4) A perpendicular bisector in an equilateral triangle is also an angle bisector (see Chapter 6).
5) $m\angle ABC = 60°$	5) The measure of each angle of an equilateral triangle is 60° (see Chapter 6).
6) $\triangle ABT \cong \triangle BCT$	6) The perpendicular bisector of an equilateral triangle splits a triangle into two congruent triangles (see Chapter 6).
7) $\overline{BT}$ equals $\frac{\sqrt{3}}{2}$ times $16\sqrt{3}$, or 24.	7) The length of the altitude of an equilateral triangle is the length of any side multiplied by $\frac{\sqrt{3}}{2}$ (see Chapter 6).
8) $\overline{OT}$ equals 24(1/3), or 8.	8) The radius of the circle inscribed in a triangle is one-third the altitude (see Chapter 6).
9) $C = 2\pi r$	9) The formula for the circumference of a circle.
10) The circumference of $\odot O$ is $C = 16\pi$.	10) Substitution (see Chapter 3).

Proof 8-15: The Circumference of an Inscribed Circle — Equilateral Triangle Proofs

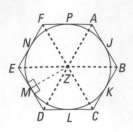

Given: ⊙Z is inscribed in regular polygon *ABCDEF*.

The radius of ⊙Z is 4.

The length of a side of polygon *ABCDEF* is s.

Find: The area of regular polygon *ABCDEF*.

Before you begin: Draw a radius for ⊙Z.

Statements	Reasons
1) ⊙Z is inscribed in regular polygon *ABCDEF*.	1) Given.
2) The radius of ⊙Z is 4.	2) Given.
3) $\overline{ZM}$ is the apothem of hexagon *ABCD*.	3) The radius of an inscribed circle in a regular polygon is also the apothem of the polygon — in this case, a regular hexagon.
4) $\overline{ZM}$ is perpendicular to $\overline{ED}$ at *M*.	4) The radius of an inscribed circle is equal to the length of the apothem of the circumscribed regular polygon.
5) ∠*ZMD* is a right angle.	5) Perpendicular lines form right angles (see Chapter 2).
6) △*MZD* is a right triangle.	6) Right triangles have a right angle (see Chapter 6).
7) m∠*EZD* = 60°	7) The sum of the central angles of a circle equals 360°, and 360°/6 = 60°.
8) m∠*MZD* = 30°	8) An apothem bisects the central angle to which it is drawn (see Chapter 5).
9) $\overline{ZM}$ equals 4.	9) All radii of a circle have the same length.
10) $4 = s\frac{\sqrt{3}}{2}$ or $s = 8/\sqrt{3}$ or approx. 4.6	10) Substitution.
11) $\overline{MD}$ = s/2 or 2.3	11) The angle opposite the 30° angle of a right triangle equals s/2.
12) $\overline{MD} = \overline{DL}, \overline{DL} = \overline{LC}, \overline{CK} = \overline{KB}, \overline{BJ} = \overline{JA}, \overline{AP} = \overline{PF}, \overline{FN} = \overline{NE}, \overline{EM} = \overline{MD}$	12) Tangents originating from the same point exterior to the circle are the same length from the exterior point to the point of tangency.

	Statements	Reasons
Proof 8-16: The Area of a Circum-scribed Regular Polygon Proof	13) $\overline{EM} + \overline{MD} = 2.3 + 2.3 = 4.6$	13) The whole is equal to the sum of its parts (see Chapter 2).
	14) The perimeter of hexagon *ABCDEF* equals 4.6(6) or 27.6 units.	14) The perimeter of a polygon is the sum of the lengths of its sides.
	15) The area of hexagon *ABCDEF* is $\frac{1}{2}(4)(27.6) = 55.2$ square units.	15) The area of a regular polygon is Area = Area $= \frac{1}{2}ap$, where *a* is the apothem and *p* is the perimeter (see Chapter 5).

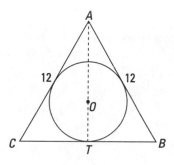

Given: $\odot O$ is inscribed in equilateral $\triangle ABC$.

$\overline{AB} = 12$

Find: The difference in the areas between the triangle and the circle.

Before you begin: Draw $\overline{AT}$ as the altitude of $\triangle ABC$.

Statements	Reasons
1) $\odot O$ is inscribed in equilateral $\triangle ABC$.	1) Given.
2) $\overline{AB} = 12$	2) Given.
3) $\overline{AB} \cong \overline{BC} \cong \overline{CA}$	3) All sides of an equilateral triangle are congruent.
4) $\overline{AC} = 12, \overline{BC} = 12$	4) Substitution.
5) $\overline{AT}$ is an altitude.	5) Drawn as an altitude.
6) $\overline{AT}$ equals $12\frac{\sqrt{3}}{2}$, or 10.39 (rounded).	6) The altitude of an equilateral triangle equals $\frac{\sqrt{3}}{2}$ times the length of any side.

(continued)

(continued)

	Statements	Reasons
Proof 8-17: The Area Differences of Inscribed/ Circumscribed Figures Proof	7) $\overline{OT}$ equals ⅓(10.39), or 3.36 (rounded).	7) The length of the radius of a circle inscribed in a equilateral triangle is one-third the length of the altitude.
	8) The area of $\odot O \approx \pi(3.46)^2 \approx 37.59$.	8) The formula for the area of a circle.
	9) The area of $\triangle ABC = \frac{1}{2}(12)(10.39) = 62.34$ square units.	9) The formula for the area of a triangle.
	10) The area of $\triangle ABC -$ the area of $\odot O = 62.35 - 37.59$, or 24.75 square units.	10) Subtraction.

Now look at a proof with a circumscribed quad. In Proof 8-19, $\odot O$ is inscribed in quadrilateral *ABCD*. Prove that the sums of the opposite sides are equal.

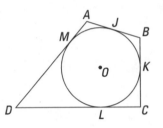

Given: $\odot O$ is inscribed in quadrilateral *ABCD*.

Prove: $\overline{AB} + \overline{CD} = \overline{BC} + \overline{DA}$

Before you begin: Label the points of tangency of the quad to the circle.

	Statements	Reasons
	1) $\odot O$ is inscribed in quadrilateral *ABCD*.	1) Given.
	2) $\overline{AJ} = \overline{AM}, \overline{BJ} = \overline{BK}, \overline{CK} = \overline{CL}, \overline{DL} = \overline{DM}$	2) Tangents originating from an exterior point are congruent from that point to the point of tangency.
	3) $\overline{AJ} + \overline{BJ} + \overline{CL} + \overline{DL} = \overline{AM} + \overline{BK} + \overline{CK} + \overline{DM}$	3) Substitution and Addition (see Chapter 3).
Proof 8-18: The Circle Inscribed in a Quadrilateral Proof	4) $\overline{AJ} + \overline{BJ} = \overline{AB}, \overline{BK} + \overline{KC} = \overline{BC}, \overline{CL} + \overline{LD} = \overline{CD}, \overline{DM} + \overline{MA} = \overline{DA}$	4) The whole equals the sum of its parts (see Chapter 2).
	5) $\overline{AB} = \overline{BC}$ and $\overline{CD} = \overline{DA}$	5) Equal segments are equal. (Put in numbers that reflect the relationships if this stuff gets too hard to follow in your head.)
	6) $\overline{AB} + \overline{CD} = \overline{BC} + \overline{DA}$	6) Substitution.

Circumscribed circles: Outside looking in

A circle is *circumscribed* about a polygon if each of the vertices of the polygon touches the circle at exactly one point.

When I first got going with geometry, I was always confused about *inscribed* and *circumscribed*. Think *in* for *inside* to remember that the inscribed circle is located inside the polygon. For circumscribed, it takes a bit more thought. How's your Latin? *Circum* means *around;* the circumscribed circle goes around the outside of the polygon and taps the polygon on the shoulder at each of its vertices.

The radius of an inscribed regular polygon is also the radius of the circumscribed circle. Figure 8-16 shows some examples of circumscribed circles.

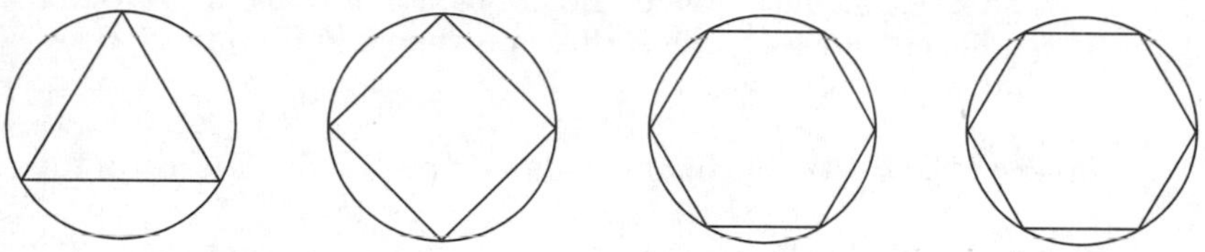

Figure 8-16:
Lots of circum-scribed circles — going *around* the polygons.

The dimensions of an inscribed polygon are restricted by the size and shape of its circumscribed circle. And all that means is this: Regardless of how many sides the inscribed polygon may have, each of its vertices must still touch the side of the circle once — the larger the radius of the circumscribed circle, the larger the size of the inscribed figure. The shape of the inscribed figure is controlled by the angles that make up the polygon.

A circle can be both circumscribed about and inscribed within any regular polygon. But all polygons don't have to be regular to be circumscribed by a circle. Certain shapes can get away with being irregular. With irregular-shaped polygons, though, it may be possible for circles to be circumscribed about but not inscribed within them. A rectangle is one such shape. A circle can be circumscribed around a rectangle, as Figure 8-17a shows, but look at Figure 8-17b to see what happens when you try to inscribe a circle within a rectangle. It just doesn't work.

Figure 8-17:
A circle can be circumscribed about a rectangle (Figure a) but not inscribed within it (Figure b).

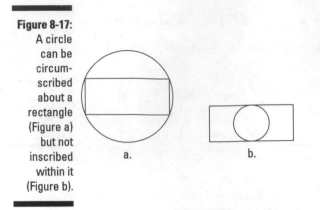

a. b.

Suppose that you dissect an inscribed polygon side by side. If you view the inscribed polygon as separate line segments as opposed to a whole figure, you can see that each side of the polygon is actually a chord of the circle. Two adjacent sides of an inscribed polygon can be viewed as two chords that form an inscribed angle.

Theorem 8-20: All sides of an inscribed polygon are chords of the circle.

Translation: All sides of an inscribed polygon are lines that go from one point on the circle to another.

Take a look at Proof 8-19. If two chords of a polygon form an inscribed angle on the circle, you can determine the degree measure of the intercepted arc.

A circle can be circumscribed about (or inscribed within, for that matter) any triangle. But for polygons with more than three sides, a restriction with the polygon's opposite angles comes into play: For the polygon to be circumscribed by a circle, the opposite angles of the inscribed polygon must be supplementary (see Figure 8-18).

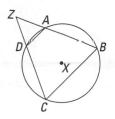

Given: ⊙X is circumscribed about quad *ABCD*.

$\overline{AD} \parallel \overline{BC}$

ZCB is a triangle.

Prove: $\overline{ZC} = \overline{ZB}$

Statements	Reasons
1) ⊙X is circumscribed about quad *ABCD*.	1) Given.
2) $\overline{AD} \parallel \overline{BC}$	2) Given.
3) m$\overset{\frown}{CDA}$ = m$\overset{\frown}{DAB}$	3) Parallel lines intercept parallel arcs.
4) m∠B = ½m$\overset{\frown}{CDA}$	4) The measure of an inscribed angle is one-half the measure of its intercepted arc.
5) m∠C = ½m$\overset{\frown}{DAB}$	5) Same as #4.
6) ½m$\overset{\frown}{CDA}$ = ½m$\overset{\frown}{DAB}$	6) Halves of equals are equal (see Chapter 3).
7) m∠B = m∠C	7) Substitution (see Chapter 3).
8) $\overline{ZC} = \overline{ZB}$	8) Opposite sides of congruent angles of a triangle are equal (see Chapter 6).

Proof 0-19:
The Inscribed Angle — Intercepted Arc of a Circle Proof

Figure 8-18:
When a polygon is inscribed in a circle, the opposite angles of the polygon are supplementary.

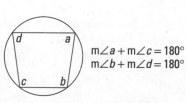

m∠a + m∠c = 180°
m∠b + m∠d = 180°

Theorem 8-21: If a circle is circumscribed about a polygon, then the interior polygon is inscribed in the circle. So it can be referred to as an *inscribed polygon*. If the inscribed polygon is a quadrilateral, the opposite angles of the inscribed quadrilateral are supplementary (refer back to Figure 8-18).

Translation: If a regular polygon is inside a circle, and the opposite angles of this polygon are added together, the resulting measure is equal to 180 degrees.

Take a look at Proof 8-20. It shows that any polygon inscribed within a circle has supplementary opposite angles.

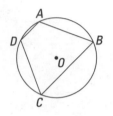

Given: $\odot O$ is circumscribed about quad *ABCD*.

Prove: $m\angle A + m\angle C = m\angle B + m\angle D$

Statements	Reasons
1) $\odot O$ is circumscribed about quad *ABCD*.	1) Given.
2) $\odot O = 360°$	2) The sum of the degree measures of the arcs in a circle equals 360°.
3) $m\angle A + m\angle C + m\angle B + m\angle D = 360°$	3) The total sum measure of the interior angles of quad *ABCD* is 360°.
4) $m\widehat{ABC} = (2)(m\angle D)$	4) The measure of the arc intercepted by an inscribed angle is twice the measure of the inscribed angle.
5) $m\widehat{ADC} = 360 - (2)(m\angle D)$	5) Subtraction.
6) $m\angle B = \frac{1}{2}(360 - (2)(m\angle D))$ or $180 - m\angle D$	6) The measure of an inscribed angle is one-half the measure of its intercepted arc.
7) $m\widehat{BCD} = (2)(m\angle A)$	7) Same as #4.
8) $m\widehat{BAD} = 360 - (2)(m\angle A)$	8) Same as #5.
9) $m\angle C = \frac{1}{2}(360 - (2)(m\angle A))$ or $180 - m\angle A$	9) Same as #6.

Statements	Reasons
10) $\widehat{mCDA} = (2)(m\angle B)$	10) Same as #4.
11) $\widehat{mCBA} = 360 - (2)(m\angle B)$	11) Same as #5
12) $m\angle D = \frac{1}{2}(360 - (2)(m\angle B))$ or $180 - m\angle B$	12) Same as #6.
13) $\widehat{mDAB} = (2)(m\angle C)$	13) Same as #4.
14) $\widehat{mDCB} = 360 - (2)(m\angle C)$	14) Same as #5
15) $m\angle A = \frac{1}{2}(360 - (2)(m\angle C))$ or $180 - m\angle C$	15) Same as #6.
16) $m\angle A + m\angle C = 360 - m\angle A - m\angle C$ $2m\angle A + 2m\angle C = 360$ $m\angle A + m\angle C = 180$	16) Addition.
17) $m\angle B + m\angle D = 360 - m\angle B - m\angle D$ $2m\angle B + 2m\angle D = 360$ $m\angle B + m\angle D = 180$	17) Addition.
18) $m\angle A + m\angle C = m\angle B + m\angle D$	18) Substitution.

Proof 8-20: The Supplementary Opposite Angles of an Inscribed Polygon Proof

Theorem 8-22: If a parallelogram is inscribed within a circle, then it is a rectangle.

Translation: If a parallelogram is inscribed within a circle, it must be a rectangle in order for all the vertices of the parallelogram to touch the circle.

Concentric circles: The center of attention

Circles that have a common center point and unequal radii are called *concentric circles*. In Figure 8-19, the measures of $\angle CXD$ and $\angle AXB$ have the same degree measure, but they don't have the same linear measure, or length. The length of arc AB is equal to $^{60}/_{360}$, or $\frac{1}{6}$, of the circle. Because the smaller circle has a smaller radius, it also has a smaller circumference. So $\frac{1}{6}$ of the length of the smaller circle isn't equal to $\frac{1}{6}$ of the length of the larger circle.

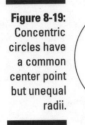

Figure 8-19:
Concentric
circles have
a common
center point
but unequal
radii.

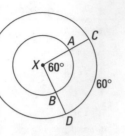

In Figure 8-19, the length of $\overline{XA}$ is not the same as the length of $\overline{XC}$ because point A and point C have two distinct locations. $\overline{XA}$ is shorter in length than $\overline{XC}$. Because $\overline{XA}$ and $\overline{XC}$ are the radii of two circles, the circle with the shorter radius is smaller than the circle with the larger radius.

Part IV
Separate but Equal? (Inequalities and Similarities)

The 5th Wave   By Rich Tennant

"I hear you think you got all the angles figured. Well, maybe you do and maybe you don't. Maybe the ratios of the lengths of corresponding sides of an equiangular right-angled triangle are equal, then again maybe they're not— let's see your equations."

In this part . . .

*I*f two quantities have the same measure, then those two quantities are equal. However, it is commonplace for things to not turn out so equitably. As a result, an inequality arises. Oh no, what do you do next? Simple. Evaluate the situation and just follow the rules! Remember not to blow things out of proportion. Quantities may not be equal but they may be similar.

Chapter 9

Inequalities (or It's Just Not Fair!)

he starter fired his pistol into the air, and with a *pop* the race began. The runners sprinted as fast as they could, gasping and grabbing for the finish line. Finally, after several grueling laps, the runner in Lane 3 broke the tape at the finish line. The crowd roared. The runner, exhausted, smiled as he waved to the stands. He won the race and captured the gold.

The runner in Lane 5 finished second, and the runner in Lane 2 finished third. The finish time of the runner who finished second was less than the one who finished third but greater than the one who won the race. And believe it or not, this is an actual example of an inequality.

In this chapter, I explain what inequality really means. And because you'll encounter inequalities in your journey through the wonderful world of geometry, I introduce some inequality proofs. They're just like any other proof, but they involve the application of some new postulates. I generally don't like to prove postulates because they're accepted as true on faith, but here I make a big exception. It's important to get an idea of the application of these postulates for when you need to apply them within more complex situations.

Inequalities 101

An *inequality* is just that — something that's not equal. If two things are not equal, then one must be either greater than or less than the other. If I have ten jellybeans, and I give you five, we both have an equal amount. But if I eat one jellybean, the situation changes depending on how you look at it. You can say that I have fewer jellybeans than you or you have more jellybeans than I.

Some geometry symbols are available to help cut down on the amount of paper it takes to write out a relationship between two numbers. Say that I start out with five jellybeans, and you have five. So 5 = 5, which reads as "five equals five." I know — *duh*. Now if I eat a jellybean, then I have to change the equation to "four is not equal to five," which is written as 4 ≠ 5. Four is also less than five, which is written as 4 < 5. Flip that and you get — if you have five jellybeans and I have four — 5 > 4, which reads as "five is greater than four."

A choice must be made regarding the status of the relationship. A statement of inequality can be true, or it can be false. If I were to say that 5 = 5, then the statement is true. But if I said that 5 ≠ 5, I'd be lying because this statement is clearly false. Only one statement can be true regarding a particular relationship at a given time.

Postulate 9-1: Any two quantities can have one of three relationships: The first quantity is equal to the second. The first quantity is greater than the second. The first quantity is less than the second.

Translation: Two quantities are equal, or one is greater than the other, or it is less than the other. You have to pick one.

I always had trouble with < and > when I first starting taking math courses. For me, an easy way to get a handle on these two symbols was to remember that the pointy part (the tip) of the arrow always points to the smaller number.

Table 9-1 provides a quick summary of the translation of inequality statements into geometric notation.

Table 9-1:	The Geometric Notation of Inequality Statements
Statement	**Notation**
a is equal to b	$a = b$
a is not equal to b	$a \neq b$
a is less than b	$a < b$
a is greater than b	$a > b$

Inequalities can travel in pairs and describe multiple relationships. For example, consider the following relationships: 8 > 7 and 3 > 2. Notice that the arrows point in the same direction. When that happens, the relationships are billed as *inequalities of the same order*. Reversing the direction of both arrows doesn't change the billing of this equation. 7 < 8 and 2 < 3 are still considered inequalities of the same order. Regardless of the direction, if the signs are both the same, then the inequalities are of the same order. Same direction means same order.

Now look at these two relationships: 3 > 2 and 5 < 7. The arrows are pointing in opposite directions. It doesn't matter whether the arrows point toward each (> <) or away from each other (< >). The important thing is that they aren't pointing in the same direction. These relationships are classified as *inequalities of the opposite order.* Opposite direction means opposite order.

Déjà vu postulates for inequalities

Consider the relationships $a > b$ and $b > c$. These are inequalities of the same order with a common member between the two. So you can apply the transitive property (see Chapter 3) to some inequalities just as easily as you can to equals or congruents. Meaning? You can make some assertions about the relationship between a and c because b is giving this information away. If $a > b$ and $b > c$, then $a > c$ (see Postulate 9-2 and Figure 9-1).

Postulate 9-2: Given three quantities, if the first is greater than the second and the second is greater than the third, then the first is greater than the third.

Translation: If $a > b$ and $b > c$, then $a > c$.

Figure 9-1:
The
transitive
property in
action: If
$\overline{AB} > \overline{BC}$ and
$\overline{BC} > \overline{CD}$,
then
$\overline{AB} > \overline{CD}$.

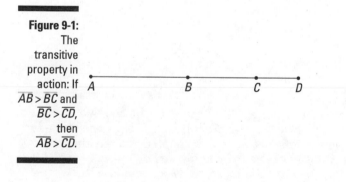

The transitive property works for angles, too (see Figure 9-2).

Figure 9-2:
If $\angle AZD > \angle 1$
and $\angle 1 > \angle 2$,
then
$\angle AZD > \angle 2$.

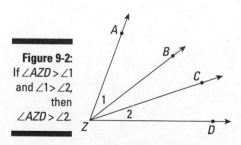

This stuff brings me to a side note: Look at Figure 9-1. $\overline{AD}$ is the length of the whole segment. $\overline{AB}$ is a part of that segment, and so are $\overline{BC}$ and $\overline{CD}$. $\overline{AB}$, $\overline{BC}$, and $\overline{CD}$ also have lengths less than that of $\overline{AD}$. So what's the point? The whole is greater than any of its parts.

Postulate 9-3: The whole quantity is greater than any one of its parts.

Translation: The whole has a greater quantity than any one of its parts.

In Figure 9-1, segments $\overline{AB}$ and $\overline{BC}$ are shorter than $\overline{AC}$. Not only is $\overline{AB} < \overline{AC}$ and $\overline{BC} < \overline{AC}$, but $\overline{AB} + \overline{BC} = \overline{AC}$. What's good for the goose is good for the gander. In this case, what's good for the segments is good for the angles. In Figure 9-3, $\angle XYZ = \angle 1 + \angle 2$, $\angle XYZ > \angle 1$, and $\angle XYZ > \angle 2$.

Figure 9-3:
An angle
whose
measure is
part of a
larger angle
has a
smaller
measure
than the
whole angle.

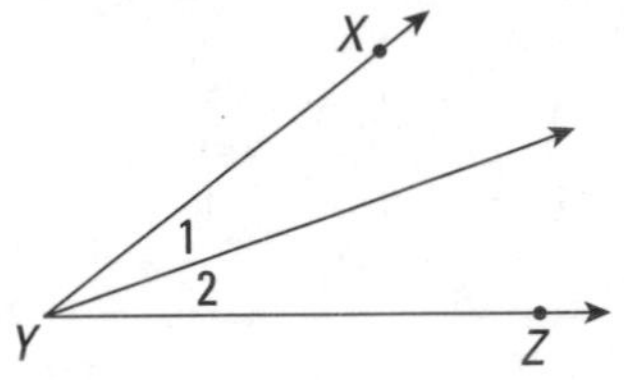

Yes, that last example was a bit simple — just to get your feet wet. How about something a little more complex? Check out Proof 9-1, which shows how to use the transitive property of inequalities. ***Note:*** I have taken some liberties in the presentation of most of the proofs in this chapter. You see, the proves in the proofs are mostly postulates and you know we don't have to prove those (see Chapter 1 for a recap). If I want to be oh-so-technical, these "proofs" are *really* just examples of how the material can be used in more complicated proofs. It's just easier to show them in proof form. It's a simplicity-comprehension thang.

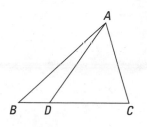

Given: m∠*BAC* > m∠*DAC* and m∠*DAC* > m∠*BAD*

Prove: m∠*BAC* > m∠*BAD*

	Statements	Reasons
Proof 9-1: The Transitive Property of Inequalities Proof	1) m∠*BAC* > m∠*DAC*	1) Given.
	2) m∠*DAC* > m∠*BAD*	2) Given.
	3) m∠*BAC* > m∠*BAD*	3) Transitive. (The first angle is greater than the second angle, the second angle is greater than the third, and so the first angle is greater than the third.)

Just as you can use the transitive property to swap values, you can also use substitution. In an inequality, you can substitute the value of any quantity for its equal. This is the essence of Postulate 9-4.

Postulate 9-4: A quantity can be substituted for another of equal value in an inequality.

Translation: You can substitute equal parts with equal parts even in an inequality.

Proof 9-2 shows how you can substitute equal parts in an inequality.

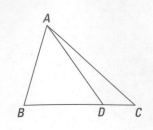

Given: $\overline{AC} > \overline{AD}$ and $\overline{AB} = \overline{AD}$

Prove: $\overline{AC} > \overline{AB}$

Proof 9-2:
The Equals
Substitution
in an
Inequality
Proof

Statements	Reasons
1) $\overline{AC} > \overline{AD}$	1) Given.
2) $\overline{AB} = \overline{AD}$	2) Given.
3) $\overline{AC} > \overline{AB}$	3) Substitution. (Substitute an equal quantity in an inequality.)

The Chaos Theory is a no-no with inequalities

You can't just throw things into the wind and see where they land. Order makes a big difference when you're dealing with inequalities. The order of the relationships, the order in which you add quantities, the order in which you subtract or multiply quantities — it all matters. The following postulates indicate how a change in the quantities of an inequality affect the order of that inequality. Get ready because in this section, I simply pile them on — postulate and proof after postulate and proof.

Postulate 9-5: If the same quantity is added to unequal quantities, the quantities are still unequal in the same order.

Translation: If $a < b$ and $c = d$, then $a + c < b + d$. With numbers, this one looks like this: If $3 < 5$ and $7 = 7$, then $3 + 7 < 5 + 7$.

Proof 9-3 shows that the same quantity added to unequal quantities maintains the status quo.

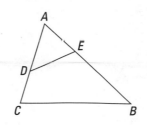

Given: $\overline{AE} = \overline{AD}$ and $\overline{BE} > \overline{DC}$

Prove: $\overline{AB} > \overline{AC}$

	Statements	Reasons
Proof 9-3: The Equals Added to Unequals Proof	1) $\overline{AE} = \overline{AD}$	1) Given.
	2) $\overline{BE} > \overline{DC}$	2) Given.
	3) $\overline{BE} + \overline{AE} > \overline{DC} + \overline{AD}$	3) If the same quantity is added to unequal quantities, their sums are unequal in the same order (see Postulate 9-5).
	4) $\overline{AB} > \overline{AC}$	4) Segment addition.

Postulate 9-6: If unequal quantities are added to unequal quantities of the same order, then the sums are unequal in the same order. You may see this postulate referred to as the Addition Property of Inequality.

Translation: If $a < b$ and $c < d$, then $a + c < b + d$. With numbers, this one looks like this: If $6 < 9$ and $2 < 3$, then $6 + 2 < 9 + 3$.

Proof 9-4 shows that relationships don't change if unequal quantities of the same order are added to an inequality.

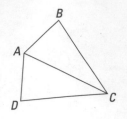

Given: m∠BCA < m∠BAC and m∠ACD < m∠DAC

Prove: m∠BCD < m∠BAD

	Statements	Reasons
Proof 9-4: The Same Order Unequals Added to Inequalities Proof	1) m∠BCA < m∠BAC	1) Given.
	2) m∠ACD < m∠DAC	2) Given.
	3) m∠BCA + m∠ACD < m∠BAC + m∠DAC	3) If unequal quantities are added to unequal quantities in the same order, their sums are unequal in the same order (see Postulate 9-6).
	4) m∠BCD < m∠BAD	4) Angle addition.

Postulate 9-7: If equal quantities are subtracted from unequal quantities, the quantities are still unequal in the same order.

Translation: If $a > b$ and $c = d$, then $a - c > b - d$. With numbers, this one looks like this: If $7 > 5$ and $4 = 4$, then $7 - 4 > 5 - 4$.

Proof 9-5 shows that if equal quantities are subtracted from unequals, the quantities are still unequal in the same order.

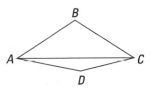

Given: m∠BCD < m∠BAD and $\overline{AB} \cong \overline{BC}$

Prove: m∠ACD < m∠DAC

	Statements	Reasons
Proof 9-5: The Equals Subtracted from Unequals Proof	1) $m\angle BCD < m\angle BAD$	1) Given.
	2) $\overline{AB} \cong \overline{BC}$	2) Given.
	3) $m\angle BCA \cong m\angle BAC$	3) In an isosceles triangle, angles opposite congruent sides are congruent (see Chapter 6).
	4) $m\angle ACD < m\angle DAC$	4) Angle subtraction.

Postulate 9-8: If unequal quantities are subtracted from equal quantities, then the differences are unequal in the opposite order.

Translation: If $c = d$ and $a < b$, then $c - a > d - b$. Here's the situation with real numbers: If $8 = 8$ and $3 < 5$, then $8 - 3 > 8 - 5$.

Proof 9-6 shows that unequals subtracted from equals still yield a difference in the opposite order.

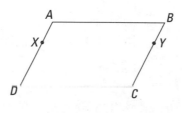

Given: *ABCD* is a parallelogram, and $\overline{CY} > \overline{AX}$.

Prove: $\overline{DX} > \overline{BY}$

	Statements	Reasons
	1) *ABCD* is a parallelogram.	1) Given.
	2) $\overline{CY} > \overline{AX}$	2) Given.
Proof 9-6: The Unequals Subtracted from Equals Proof	3) $\overline{BC} \cong \overline{AD}$	3) Opposite sides of a parallelogram are congruent.
	4) $\overline{AD} - \overline{AX} > \overline{BC} - \overline{CY}$	4) If unequals are subtracted from equals of the same order, the quantities are still unequal in the opposite order (see Postulate 9-7).
	5) $\overline{DX} > \overline{BY}$	5) Subtraction

Postulate 9-9: If unequal quantities are multiplied by the same positive number, the products are unequal in the same order.

Translation: If $a > b$ and $c = d$, then $ac > bd$, given that a, b, c, and d are all positive numbers. With real numbers, you've got this: If $8 > 7$ and $2 = 2$, then $(8)(2) > (7)(2)$.

A special case of Postulate 9-9 involves doubles of unequal quantities, which are also unequal in the same order. In other words, if $a < b$, then $2a < 2b$. Using real numbers, if $7 < 9$, then $(2)(7) < (2)(9)$.

Proof 9-7 shows that unequals multiplied by equal positive numbers are unequal in the same order, per Postulate 9-9.

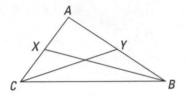

Given: $\overline{AX} < \overline{AY}$. $\overline{BX}$ and $\overline{CY}$ are medians of $\triangle ABC$.

Prove: $\overline{AC} < \overline{AB}$

Statements	Reasons
1) $\overline{AX} < \overline{AY}$	1) Given.
2) $\overline{BX}$ and $\overline{CY}$ are medians of $\triangle ABC$.	2) Given.
3) $\overline{CX} \cong \overline{AX}$	3) A median splits a side into two equal segments (see Chapter 6).
4) $\overline{AY} \cong \overline{BY}$	4) Same as #3.
5) X is the midpoint of $\overline{AC}$.	5) A median meets a side at the midpoint of that side (see Chapter 6).
6) Y is the midpoint of $\overline{AB}$.	6) Same as #5.
7) $2\overline{AX} < 2\overline{AY}$	7) Unequals multiplied by the same quantity are unequal in the same order (see Postulate 9-9).
8) $\overline{AC} < \overline{AB}$	8) Substitution (see Chapter 3).

Proof 9-7: The Unequals Multiplied by Equal Positives Proof

Postulate 9-10: If unequal quantities are multiplied by the same negative number, the results are unequal in the opposite order.

Translation: If $a < b$ and $c = d$, where c is a negative number, then $ac > bd$. Using real numbers, if $4 < 5$ and $-2 = -2$, then $(4)(-2) > (5)(-2)$, or $-8 > -10$.

Note that I don't like using negative numbers in a proof. Bad karma. (Besides, it's hard to show negative length.)

Postulate 9-11: If unequal quantities are divided by the same positive number, the quotients are unequal in the same order.

Translation: If $a > b$ and $c = d$, then $a/c > b/d$. Here's a real-number version of this one: If $7 > 4$ and $3 = 3$, then $7/3 > 4/3$.

A special case of Postulate 9-11 involves the halves of unequal quantities. Halves of unequals are unequal in the same order. In other words, if $a < b$, then $a/2 < b/2$.

Proof 9-8 shows that unequals divided by an equal positive are unequal in the same order, per Postulate 9-11.

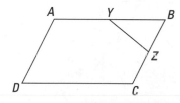

Given: $\overline{AB} > \overline{BC}$, $\overline{BZ} \cong \frac{1}{2}\overline{BC}$, $\overline{AY} \cong \frac{1}{2}\overline{AB}$

Prove: $\overline{BY} > \overline{BZ}$

Proof 9-8:
The
Unequals
Divided by
Equals Proof

Statements	Reasons
1) $\overline{AB} > \overline{BC}$	1) Given.
2) $\overline{BZ} \cong \frac{1}{2}\overline{BC}$	2) Given.
3) $\overline{AY} \cong \frac{1}{2}\overline{AB}$	3) Given.
4) $\overline{BY} + \overline{AY} = \overline{AB}$	4) Y is the midpoint of $\overline{AB}$, and a whole is the sum of its parts.
5) $\overline{BY} = \frac{1}{2}\overline{AB}$	5) Substitution (see Chapter 3).
6) $\overline{AB}/2 > \overline{BC}/2$	6) Halves of unequals are unequal in the same order.
7) $\overline{BY} > \overline{BZ}$	7) Substitution (see Chapter 3).

Postulate 9-12: If unequal quantities are divided by equal negative quantities, the quotients are unequal in the opposite order.

Translation: If $a > b$ and $c = d$, where c is a negative number, then $ac < bd$. Here's a real-life example: If $6 > 3$ and $-3 = -3$, then $6/-3 < 3/-3$, or $-2 < -1$.

Postulate 9-13: Equal positive integral powers and equal positive integral roots of unequal positive quantities are unequal in the same order.

Translation: If $a > b$, where a and b are positive integers, then $a^n > b^n$ (where n is a positive integer) and $\sqrt{a} > \sqrt{b}$. In other words, if $4 > 3$, then $4^2 > 3^2$ or $\sqrt{4} > \sqrt{3}$. You get the idea, so I'll spare you the proof on this one. (You're welcome.)

Triangle Inequality

When I fly to Puerto Rico, I like to take a direct flight. But sometimes, especially during the off-season, I have to take a connecting flight. I *hate* connecting flights. They always take longer. Besides the layover time, the flying distance is actually longer. I have to fly out of my way to get to where I want to go. On my last trip, I was lucky enough to get a direct flight out, but I had to take a connecting flight back. My route formed a triangle (see Figure 9-4).

The direct route from Washington, D.C. to San Juan was shorter than the combined length of the return trip from San Juan to Atlanta and then from Atlanta back to Washington, D.C. Because I was traveling a triangle route, the length of the longest side of this triangle (Washington, D.C. to San Juan) was greater than the difference of the lengths of the two shorter sides (San Juan to Atlanta and Atlanta to Washington, D.C.) and less than the sum of these two lengths. Simplified, it looks something like this: $s - a < w < s + a$, where s is the distance from San Juan to Atlanta, a is the distance from Atlanta to

Washington, D.C., and w is the distance from Washington, D.C., to San Juan. This stuff is actually a theorem, known as the Triangle Inequality Theorem, and it can take one of two forms.

Theorem 9-1: The sum of the lengths of any two sides of a triangle is greater than the length of the third side.

Translation: Add the lengths of any two sides of a triangle, and the result is greater in length than the third side.

Theorem 9-2: The length of the longest side of a triangle is greater than the difference of the two other sides and less than the sum of those two sides.

Translation: The length of the longest side of a triangle (c) is greater than the difference of the other two sides ($a - b$) and less than the sum of those two sides ($a + b$). Put it all together, and you get $a - b < c < a + b$.

Proof 9-9 shows that the sum of two lengths of a triangle is greater than the length of the third side.

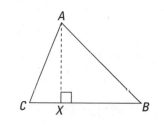

Given: *ABC* is a triangle.

Prove: $\overline{AC} + \overline{AB} > \overline{BC}$

	Statements		Reasons
1)	*ABC* is a triangle.	1)	Given.
2)	$\overline{AX}$ is perpendicular to $\overline{BC}$.	2)	Drawn as perpendicular. A perpendicular is the shortest distance from a point to a line (see Theorem 2-9).
3)	$\overline{AC} > \overline{CX}$ and $\overline{AB} > \overline{BX}$	3)	The hypotenuse is the longest side of a right triangle. Triangles ACX and ABX are right triangles formed by the perpendiclar (see Chapter 6).
4)	$\overline{AC} + \overline{AB} > \overline{CX} + \overline{BX}$	4)	If unequals are added to an inequality in the same order, the sums are unequal in the same order.
5)	$\overline{AC} + \overline{AB} > \overline{BC}$	5)	Substitution.

Proof 9-9:
The Triangle
Side Length
Proof

Re-examine the route map for my trip. The legs of my trip are different lengths, which means that two sides of my triangle are unequal. If two sides of a triangle are equal (congruent), then their opposite angles are equal (congruent), (see Theorem 6-9). I'm sure you already see the writing on the wall with this one: If the lengths of two sides of a triangle are unequal, then the measures of the angles opposite those sides are also unequal — with the larger angle being opposite the longer side. Makes sense to me! Theorem 9-3 sums up this concept nicely.

Theorem 9-3: If the lengths of two sides of a triangle are unequal, the measures of the angles opposite those sides are unequal. The largest angle of a triangle is opposite the longest side.

Translation: Unequal sides map to unequal opposite angles. The largest angle is opposite the longest side of a triangle.

Now how about a visual? In Figure 9-5, $\overline{XY}$ is longer than $\overline{YZ}$. The measure of angle Z, which is opposite $\overline{XY}$, is greater than the measure of the angle X, which is opposite $\overline{YZ}$.

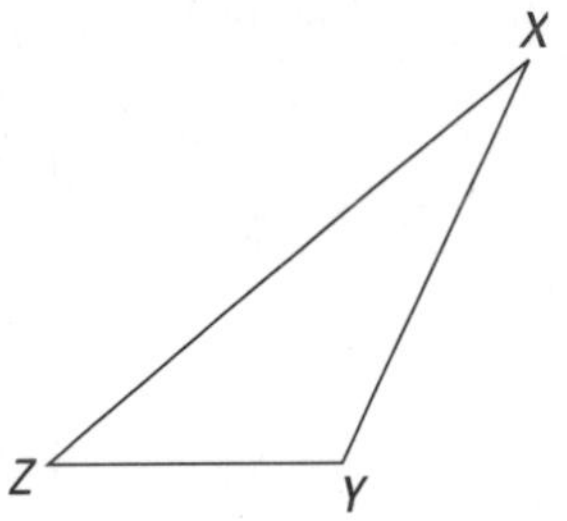

Figure 9-5:
Two unequal sides have unequal opposite angles.

What happens if you have two triangles with two corresponding congruent sides? The triangle with the largest included angle has the greater third side. For example, look at the two triangles in Figure 9-6. They have two corresponding congruent sides. Angle *A* of triangle *ABC* is larger than angle *X* of triangle *XYZ*, so side $\overline{BC}$ has to be longer than $\overline{YZ}$.

You can also look at this concept from a new angle. Or, more precisely, *from the angles.* Flip the wording around — *sides* for *angles* — and you get this: If the measures of two angles of a triangle are unequal, the lengths of the sides opposite those angles are also unequal. The longer side is opposite the larger angle.

Figure 9-6:
If two triangles have two corresponding congruent sides, then the triangle with the largest included angle has the largest third side.

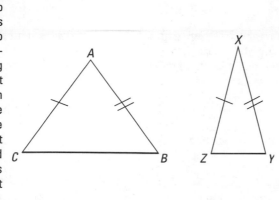

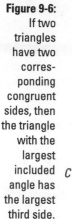

Theorem 9-4: If the measures of two angles of a triangle are unequal, the lengths of the sides opposite those angles are unequal. The longest side of a triangle is opposite the largest angle.

Translation: Unequal angles map to unequal opposite sides. The longest side is opposite the largest angle of a triangle.

In Figure 9-7, the measure of $\angle F$ is greater than the measure of $\angle E$. So the length of side $\overline{EG}$ is greater than the length of $\overline{FG}$.

Figure 9-7:
Two unequal angles have unequal opposite sides.

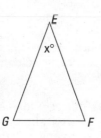

Food for thought: The sum measure of $\angle F$ and $\angle E$ is equal to the measure of an exterior angle supplementary to $\angle G$. (This info might come in handy at some point, so I thought I'd bring it up.)

Now if two triangles have two congruent corresponding sides, and the third side doesn't match any other side size-wise, then the triangle with the greater third side has the greater opposite angle. In Figure 9-8, $\overline{CB} > \overline{GF}$, so $\angle A > \angle E$.

Figure 9-8:
If two triangles have two corresponding congruent sides, then the triangle with the largest third side has the largest opposite angle.

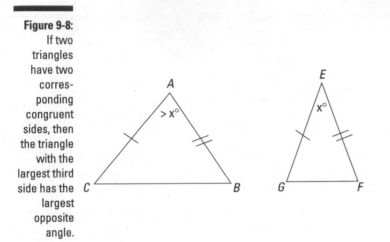

Circle Inequality

The inequalities of a circle generally relate to its lines, angles, and arcs. I guess that means they apply to just about any aspect of a circle that can be compared and expressed as an inequality.

I'm sure this one will strike a *chord* with you: If two different lines are drawn from a common point on a circle to other (different) points on the circle, these lines (also known as chords) are judged by their distance from the center of the circle. The farther from the center of the circle a chord is, the shorter the chord's length. The flip side of that is also true: The longer the chord, the closer it is to the circle's center.

The largest chord in a circle is the diameter (see Chapter 8).

I'm sure you're wondering how this information could possibly be of use. For a start, check out Figure 9-9. In that figure, ⊙P contains chords $\overline{AB}$ and $\overline{XY}$. $\overline{PQ}$ extends from the center of the circle to $\overline{XY}$, and $\overline{PM}$ extends from the center of the circle to $\overline{AB}$. $\overline{XY} > \overline{AB}$, so $\overline{PQ} < \overline{PM}$. Why? Because $\overline{XY}$ is closer to the center of the circle and $\overline{AB}$ is farther from the center.

Figure 9-9:
A chord that's closer to the center of a circle is longer than a chord that's farther away from the center.

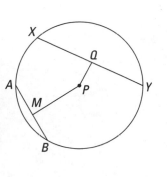

Theorem 9-5: The closer a chord is to the center of a circle, the longer the chord is.

Translation: A chord is longer when it is closer to the center of a circle.

Theorem 9-6: The farther a chord is from the center of a circle, the shorter the chord is.

Translation: A chord is shorter when it is farther from the center of a circle.

Proof 9-10 uses Theorems 9-5 and 9-6 to show that chords farther from the center of a circle are shorter and that chords closer to the center of a circle are longer.

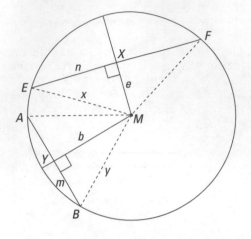

Given: $\overline{EF}$ and $\overline{AB}$ are chords of $\odot M$, $\overline{MX} \perp \overline{EF}$. $\overline{MY} \perp \overline{AB}$, and $\overline{MX} < \overline{MY}$.

Prove: $\overline{AB} < \overline{EF}$

Before you begin: Draw radii $\overline{ME}, \overline{MF}, \overline{MB},$ and $\overline{MA}$.

Statements	Reasons
1) $\overline{EF}$ and $\overline{AB}$ are chords of $\odot M$.	1) Given.
2) $\overline{MX} \perp \overline{EF}$	2) Given.
3) $\overline{MY} \perp \overline{AB}$	3) Given.
4) $\overline{ME}, \overline{MF}, \overline{MB},$ and $\overline{MA}$ are radii of $\odot M$.	4) Drawn as radii.
5) $\overline{MX}$ is a perpendicular bisector of $\overline{EF}$.	5) A radius perpendicular to a chord bisects that chord (see Chapter 8).
6) $\overline{MY}$ is a perpendicular bisector of $\overline{AB}$.	6) Same as #6.
7) $\triangle MYB$ and $\triangle MXE$ are right triangles.	7) Right triangles contain a right angle (see Chapter 6).
8) $x^2 = e^2 + n^2$ and $y^2 = m^2 + b^2$	8) The lengths of the sides of a right triangle comply with the Pythagorean Theorem (see Chapter 6).
9) $\overline{MX} < \overline{MY}$	9) Given.
10) $e < b$	10) Substitution (see Chapter 3).
11) $e^2 < b^2$	11) Squares of inequalities are unequal in the same order.
12) $-e^2 > -b^2$	12) Unequals multiplied by the same negative number (in this case, -1) are unequal in the opposite order.

13) $y \cong x$	13) All radii of a given circle are congruent (see Chapter 8).
14) $x^2 - e^2 > x^2 - b^2$ or $x^2 + (-e^2) > x^2 + (-b^2)$	14) Equal quantities added to inequalities result in unequals in the same order. (Note: I am adding a positive number (x^2) to a negative number ($-e^2$ or $-b^2$), see #12 above.)
15) $n^2 > m^2$	15) Substitution (see Chapter 13).
16) $n > m$	16) Square roots of inequalities are unequal in the same order.
17) $2n > 2m$	17) Unequals multiplied by the same number are unequal in the same order.
18) $\overline{EF} > \overline{AB}$	18) Substitution (see #5 and #6 above, as well as Chapter 3).
19) $\overline{AB} < \overline{EF}$	19) Symmetric (see Chapter 3).

Proof 9-10:
The Chord Location Proof

To sum up, if you want to prove that two chords are unequal, you have to show one of two things: First, show that the chords are from the same circle or from equal circles. Or, show that their arcs are unequal or that the arcs are not equidistant from the center of the circle.

You can also use chords to gauge the relative relationship of the minor arcs that they cut off from the circle. The longer the chord, the closer to the center of the circle the chord lies. The closer to the center of the circle the chord lies, the greater the size of its minor arc. Go for the visual — look at Figure 9-10. Chord $\overline{AB} > \overline{MN}$, and $\overset{\frown}{AB} > \overset{\frown}{MN}$.

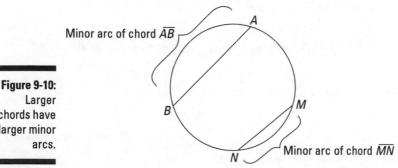

Figure 9-10:
Larger chords have larger minor arcs.

Minor arc of chord $\overline{AB}$

Minor arc of chord $\overline{MN}$

Equal chords cut off equal arcs (see Postulate 8-2).

If the minor arc of a circle gets larger the longer a chord is (up to a point of course), then that space has to come from somewhere. It's taken from the major arc: the larger the minor, the smaller the major. There are so many ways to look at this stuff. For example, shorter chords have smaller minor arcs but larger major arcs. You get the idea.

Theorem 9-7: In the same circle or equal circles, the larger minor arc has the longer chord.

Translation: Under equal circle circumstances, the larger minor arc has the longer chord.

Theorem 9-8: In the same circle or equal circles, the smaller minor arc has the shorter chord.

Translation: Under equal circle circumstances, the smaller minor arc has the shorter chord.

Proof 9-11 shows that the relative lengths of chords and the size of their minor arcs are related.

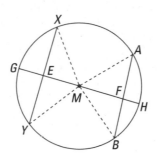

Given: $\overline{XY}$ and $\overline{AB}$. are chords of $\odot M$, and $m\widehat{AGB} < m\widehat{XHY}$.

Prove: $m\overline{AB} < m\overline{XY}$

Before you begin: Add radii $\overline{MX}, \overline{MY}, \overline{MB},$ and $\overline{MA}$.

Statements	Reasons
1) $\overline{XY}$ and $\overline{AB}$ are chords of $\odot M$.	1) Given.
2) $\overline{MX}, \overline{MY}, \overline{MB},$ and $\overline{MA}$ are radii.	2) Drawn as radii.
3) $\overline{MX} \cong \overline{MY} \cong \overline{MB} \cong \overline{MA}$	3) Radii of the same circle are congruent.
4) $m\widehat{AGB} < m\widehat{XHY}$	4) Given. (These are major arcs.)

Statements	Reasons
5) $-\text{m}\overset{\frown}{AGB} > -\text{m}\overset{\frown}{XHY}$	5) Unequals multiplied by the same negative number are unequal in the opposite order. (Note: I multiplied each side by –1.)
6) $C = \text{m}\overset{\frown}{AGB} + \text{m}\overset{\frown}{AB}$	6) The circumference of a circle equals the sum of its unique minor and major arcs (see Chapter 8).
7) $C = \text{m}\overset{\frown}{XHY} + \text{m}\overset{\frown}{XY}$	7) Same as #6.
8) $C + (-\text{m}\overset{\frown}{AGB}) > C + (-\text{m}\overset{\frown}{XHY})$	8) Equals added to unequals result in unequals in the same order. (C is the circumference of the circle. C is the equal amount added to $-\text{m}\overset{\frown}{AGB}$ and $-\text{m}\overset{\frown}{XHY}$.)
9) $\text{m}\overset{\frown}{AB} < \text{m}\overset{\frown}{XY}$	9) Substitution.

Proof 9-11:
The Chord
and Arc
Length
Proof

Central angles also have an effect on arc size. In the same circle or equal circles, the larger the central angle, the larger its intercepted arc. The smaller the central angle, the smaller its intercepted arc.

Theorem 9-9: In the same circle or equal circles, the central angle with the greater measure has the greater arc.

Translation: In the same circle or equal circles, the greater the central angle, the greater its arc is.

Theorem 9-10: In the same circle or equal circles, the central angle with the lesser measure has the smaller arc.

Translation: In the same circle or equal circles, the smaller the central angle, the smaller its arc is.

Proof 9-12 shows that larger central angles have greater arcs and that smaller central angles have lesser arcs.

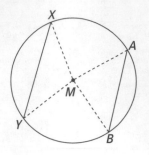

Given: $\overline{XY}$ and $\overline{AB}$ are chords of $\odot M$, and $\overline{AB} < \overline{XY}$.

Prove: $\text{m}\widehat{AB} < \text{m}\widehat{XY}$

Before you begin: Add radii $\overline{MX}, \overline{MY}, \overline{MB}$, and $\overline{MA}$.

Statements	Reasons
1) $\overline{XY}$ and $\overline{AB}$ are chords of $\odot M$.	1) Given.
2) $\overline{MX}, \overline{MY}, \overline{MB}$, and $\overline{MA}$ are radii of $\odot M$.	2) Drawn as radii of $\odot M$.
3) $\overline{AB} < \overline{XY}$	3) Given.
4) $\overline{MX} \cong \overline{MY} \cong \overline{MB} \cong \overline{MA}$	4) Radii of the same circle are congruent.
5) $\angle AMB < \angle AMY$	5) If two triangles have two corresponding congruent sides, then the triangle with the larger third side ($\overline{XY}$) has the larger included angle.
6) $\text{m}\widehat{AB} < \text{m}\widehat{XY}$	6) In the same circle or equal circles, the central angle with the lesser measure has the smaller arc.

Proof 9-12:
The Central
Angle-Arc
Proof

To sum up, if you want to prove that two arcs are unequal, you have to show one of two things: First, show that the two arcs are from the same circle or from equal circles. Or, show that their central angles are unequal or that their chords are unequal.

Chapter 10

Similarity Is the Sincerest Form of Flattery

*L*ast month my Aunt Suzie helped me redecorate my apartment. Using her background in architecture and interior design, she drew a floor plan of each of the rooms. She didn't have — nor did she need — a 600-square-foot piece of paper. She reduced the size of the floor plan so that it accurately represented the features of my apartment in a relative size. Her drawing was a scaled-down version of reality, just as dollhouses and train sets can be miniatures of larger, real-world objects.

In this chapter, you get to shrink objects in proportion. After you get done here, you'll be able to find the measurements of large objects, like the Empire State Building or the Statue of Liberty, using similar triangles and proportions and no ruler. And your payoff? You'll learn what the odds at a horse race really mean.

Relating to Ratios

Say you put ten marbles in a bag. Two are black, and the rest are white. How does the number of black marbles compare to the number of white marbles? The first thing you need to do is put all the info into numbers: two black marbles and eight white marbles (see Figure 10-1).

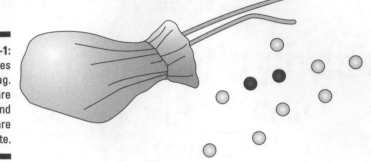

Figure 10-1:
Ten marbles
are in a bag.
Two are
black, and
the rest are
white.

You can compare these numbers by using a ratio. A *ratio* is the quotient of two numbers, where the denominator is not zero. In general form, a ratio can be expressed as $a/b\,(b \neq 0)$. The ratio in the marble example is 2/8, which reads as "2 is to 8." But this is only one way you can express a ratio. You can also write a ratio by using a colon, like 2:8. You can use the word *to,* as in 2 to 8. There's the decimal — .25 — and the percent — 25%.

The *traditional ratio* compares two numbers, but you can also use a ratio to compare more than two numbers. A ratio that compares more than two numbers is called an *extended ratio* or a *continued ratio*. It goes above and beyond the traditional ratio and can look something like this: 2:7:8. Three comparisons are contained within this one string of numbers and colons: 2/7, 2/8, and 7/8.

Converting to the same units of measure

A ratio is a unitless comparison, so when you compare, say, the number of black marbles to white marbles, the ratio makes no mention of the words *black, white,* or *marbles.* I don't want you to get the wrong idea, though. You can't just throw any numbers together to form a ratio. They have to be of the same unit measure *before* you put them together. The units of a ratio must be *commensurable,* meaning that you can convert them to the same unit of measure. For example, say you want to compare 6 inches and 3 feet. A ratio of 6/3 would be wrong as a comparison because you first have to convert the 3 feet into their comparable number of inches.

In this section, I walk you through the conversion process. I learned this method for doing conversions in my high school chemistry class, and it's still the easiest way I know to do them. The best thing to do is write everything down and lay it all out. If you do things in your head, you can end up skipping steps without ever realizing it, but if you put everything down on paper, it's all there in black and white.

When doing conversions, make sure that you write down the units for each number in the equation. Doing so ensures that you end with the units you want.

1. **Start with the quantity that has the units you want to get rid of.**

   If you want to compare 6 inches and 3 feet and you want to work in inches, for example, then you want to start with the quantity of 3 feet.

2. **Set up the equation.**

   In this example, you need to convert feet to inches. Put a 1 under 3 feet because you know that 3/1 is the same as 3.

   This step just helps you keep the terms on the level of the fraction that they're supposed to be on. That is, it's easier if you compare one double-decker quantity to another double-decker quantity. If you don't feel that you need to put in the 1, you don't have to. But remember that the 3 feet is on the top level, with an implied 1 on the bottom level.

   $$\frac{3 \text{ feet}}{1}$$

3. **Introduce the conversion into the equation.**

   Next, multiply the 3 feet (over 1 fraction) by 1. This isn't any ordinary number 1 that you multiply by, though. It's the unit's equivalent of 1. You use this equivalent to change the units of 3 feet to inches.

   In this example, you know that 1 foot equals 12 inches and you want the units of feet to cancel out. For the units to cancel out, the unit of feet must be on the top of one fraction and on the bottom of another fraction. So all you have to do is multiply 3 feet by the appropriate fraction:

   $$\frac{3 \text{ feet}}{1} \times \frac{12 \text{ inches}}{1 \text{ foot}} = 36 \text{ inches}$$

4. **Resolving the equation.**

   The units of feet cancel each other out, and you're left with 3 times 12 inches. The product is 36 inches. So 3 feet equals 36 inches.

You can now use the converted number in a ratio because it matches the units of the number you want to compare with. In the example of 6 inches and 3 feet, you can now correctly write the ratio as 6/36 or 1/6.

Using ratios in geometry

Because a ratio is a comparison, you can use it to compare any commensurable quantities, including the lengths of the sides of a triangle, the measures of angles, or line segments.

For example, take a look at Figure 10-2. Side $\overline{AB}$ of $\triangle ABC$ is 10 inches long, and side $\overline{BC}$ is 8 inches long. The ratio of the lengths of these two sides is 10/8. $\overline{AB}$ is 1.25 times the length of $\overline{BC}$.

Figure 10-2:
You can use ratios to compare the lengths of the sides of a triangle.

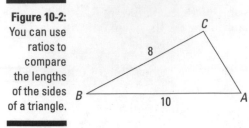

Now look at Figure 10-3. The measure of $\angle BAC$ to $\angle 1$ is 90/45, or 2. The measure of $\angle BAC$ is twice that of $\angle 1$.

Figure 10-3:
You can use ratios to compare the measures of angles.

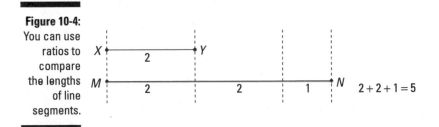

One more: In Figure 10-4, $\overline{MN}$ equals 5 and $\overline{XY}$ equals 2. The ratio of the lengths of these two line segments is 5/2. $\overline{MN}$ is 2.5 times the length of $\overline{XY}$.

Figure 10-4:
You can use ratios to compare the lengths of line segments.

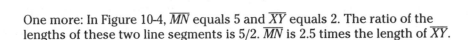

Comparing with Rates

What happens when you want to compare two items but you can't change the units of measure so that they're alike? I'm sure you've heard the saying, "You can't compare apples to oranges." That saying basically sums it up. You can't form a ratio to compare quantities if there's no common unit between them. These quantities are called *incommensurable*. But that doesn't mean the information they give you is useless. Two commensurable quantities form a ratio, but two incommensurable quantities form a *rate*.

Comparing miles per gallon

Suppose that you're driving on the highway to visit Mother Dear, and you're trying to decide whether you should stop for gas before you get to her house. You have a half-tank of gas. You know that your truck's gas tank holds 20 gallons. Half of 20 is 10, so you have 10 gallons of gas in the tank. You just passed a sign that indicates you have to travel 180 more miles until you get to your destination. You have 10 gallons of gas and 180 miles to go. Now you need a bit more information — like how many miles to the gallon your truck uses — before you can decide whether to stop for gas before getting to Mom's place.

Miles to the gallon, or miles per gallon, is an example of units that are incommensurable. An incommensurable comparison stills yields useful information, though, especially if you don't want to run out of gas and end up walking to your mom's house.

OK, your gas-guzzling truck travels 18 miles per gallon. You can write this rate as 18 mpg. So, if you have 10 gallons of gas, and your truck travels 18 miles per gallon, you can multiply these quantities by using the following equation:

$$\frac{10 \text{ gallons}}{1} \times \frac{18 \text{ miles}}{1 \text{ gallon}} = 180 \text{ miles}$$

Set up your equation beginning with the 10 gallons of gasoline that are in your truck. Place a 1 under the 10 gallons so that you have a fraction to multiply with. Next, go for the rate, 18 miles per 1 gallon. Ten gallons as the gallons unit is on the top of the fraction. You want miles as your ending unit, so you must have gallons on the bottom of the second fraction. Once you set it up this way, the gallons units cancel out. And, you know how far you can travel on your tank of gasoline — 180 miles.

You have 180 miles to travel to your mom's, and you have 180 miles' worth of gas in the tank. A bit too close for comfort, I'd think. If you don't like driving on fumes, stop for gas before getting to your mother's house.

Comparing miles per hour

Miles per hour is another example of incomensurable quantities. Say that the speed limit on the highway to your mother's house is 65 miles per hour, or 65 mph. You have 180 miles to travel to get to your destination. You set your cruise contol at 65 mph. How long will it take you to get there? Check out this formula to find out:

$$\frac{180 \text{ miles}}{1} \times \frac{1 \text{ hour}}{65 \text{ miles}} \approx 2.77 \text{ hours}$$

Notice in the conversion equation that *1 hour* is on the top and *65 miles* is on the bottom. What you need to do here is ask, "What unit do I start with?" and "What unit do I want to end with?" You want the unit that you *need* to be the only unit left in the equation when you're done with the conversion. A unit is removed when one quantity with this unit is on the top of a fraction and another quantity with this unit is on the bottom of another fraction. At this point, the units cancel each other out.

In the miles per hour example, you want to start with how far you need to travel, and you want to end with how long it'll take you to get there. Let the units guide how your equation gets set up. That way, you don't have to think about whether you need to multiply or divide. You just have to set up your equation so that the only unit left is, in this case, hours. The math that needs to be done will already be mapped out for for you. In this case, when you put the conversion equation together, it tells you to multiply 180 times 1 across the top of the equation and to multiply 1 times 65 across the bottom. The result is 180/65. You can call Mom and let her know that from your current location, you'll be at her house in approximately 2.77 hours.

Comparing cost per unit

The rate information that I find particularly useful is the cost per unit information that is below every product in the supermarket. I'm one of those grocery geeks who compares the value of one product to the value of another.

Suppose, for example, that one bottle of dishwashing soap costs $1.29 and another bottle costs 99 cents. By just looking at the price, you may think that the 99-cent bottle is a better deal because it's cheaper. But by looking a little closer, you can see that what seemed like a deal isn't one. The bottle that costs 99 cents is only 10 fluid ounces (abbreviated as 10 fl. oz.), but the bottle that costs $1.29 is 14 fluid ounces. The unit price allows you to compare the cost of the two products based on the cost per fluid ounce. In this case, the 10 fl. oz. bottle costs 9.9 cents per fluid ounce, and the 14 fl. oz. bottle costs 9.2 cents per fluid ounce. The 14 fl. oz. bottle is actually a better deal (see Figure 10-5).

Figure 10-5:
If one bottle of dish-washing liquid costs less than another, the cheaper one isn't necessarily a better deal. Compare the cost per unit rate.

$1.29

$0.99

14 oz.

10 oz.

Using rates in geometry

Rates are very useful in geometry. They are particularly useful in the formation of ratios that are used in equations called proportions. Proportions are, well, covered next...read on!

Keeping It in Proportion

A *proportion* is an equation stating that two ratios or two rates are equal to each other. A proportion can take on several different appearances. One is $a:b = c:d$. More commonly in geometry, fractions are used and take the following general form:

$$\frac{a}{b} = \frac{c}{d}$$

This proportion reads as "*a* is to *b* as *c* is to *d*." The terms of a proportion are referred to by their location in the proportion. Look at the sentence used to describe the proportion in words. It's the best way to understand the naming thing. *a* is the first quantity in the sentence "*a* is to *b* as *c* is to *d*," and it's the first term of the proportion. So you refer to *a* as the *first proportional*. *b* is the second term in the sentence and is the *second proportional*. *c* is the third term in the sentence and is the *third proportional*. The *fourth proportional* is *d*, the last term in the proportion. Figure 10-6 gives you a visual of the names.

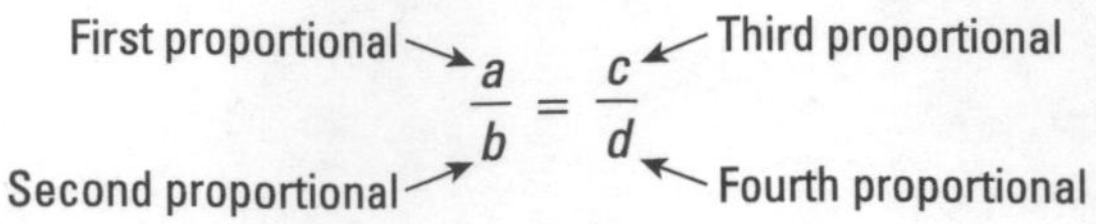

Figure 10-6:
The quantities of a proportion are named by their location in the proportion.

First proportional →

Second proportional →

$$\frac{a}{b} = \frac{c}{d}$$

← Third proportional

← Fourth proportional

The first and fourth quantities of the proportion are the outermost quantities in the proportion. For this reason, they're called the *extremes*. The second and third quantities of the proportion are referred to as the *means*. These quantities are in the middle of the equation, and the word *means* roughly translates to "midway." Figure 10-7 gives you a visual of these terms. In the figure, they're identified in both of the ways you can view a proportion.

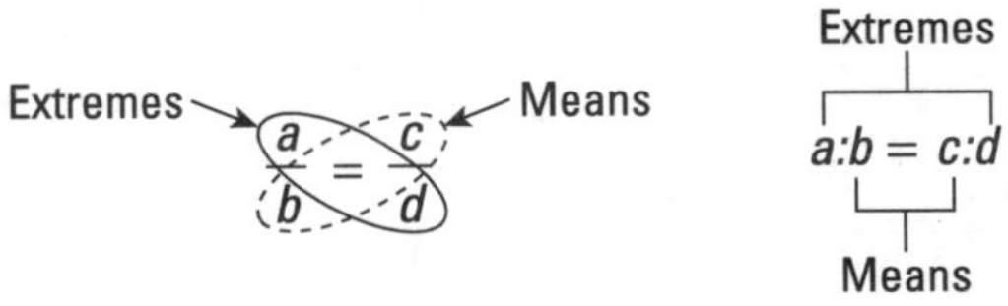

Figure 10-7:
The means and extremes of a proportion are related to the location of a quantity in the proportion.

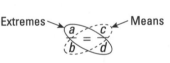

Extremes → | ← Means

Extremes

$$a{:}b = c{:}d$$

Means

To determine whether a proportion is true, you multiply the two mean quantities together and the two extreme quantities together. These two products will be equal if the proportion is true. Here's a good method to use: Lightly draw an X across the proportion. Numbers that fall on the same line are multiplied together. Multiply the *numerator* (the top number) of one fraction by the *denominator* (the bottom number) of the other fraction. When quantities are multiplied across the equal sign like this, the products are called *cross-products*, and the method of obtaining these products is called *cross-multiplication*.

In Figure 10-8, for example, I drew an X, or cross, on my proportion to determine which quantities to multiply together when I cross-multiply. In this case, 7 and 2 are multiplied together, and 14 and 1 are multiplied together.

The cross-products are both 14, and the fractions are reduced to a result that looks like this: 14 = 14. This equality statement is true, so the equation in Figure 10-8 is a proportion.

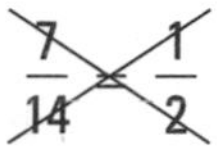

Figure 10-8:
When cross-multiplying, draw an X on your equation and multiply together the quantities that are on the same line.

$$\frac{7}{14} \diagdown\!\!\!\!\diagup \frac{1}{2}$$

Here is a different scenario. The quantities

$$\frac{6}{12} = \frac{3}{4}$$

are presented as a proportion. If this statement truly is a proportion, then 6 multiplied by 4 should be equal to 12 multiplied by 3:

$$6 \times 4 = 12 \times 3$$
$$24 = 36 \quad \text{(This is not true.)}$$
$$24 \neq 36 \quad \text{(This is true.)}$$

Because it's not true that 24 = 36, then

$$\frac{6}{12} = \frac{3}{4}$$

is not a proportion.

Now look at this proportion:

$$\frac{6}{12} = \frac{3}{6}$$

Determine the cross-products. You get 36 = 36. The two quantities obtained from the cross-products of the proportion are equal, so

$$\frac{6}{12} = \frac{3}{6}$$

is a proportion. The Means-Extremes Property states just that: In a proportion, the product of the means equals the product of the extremes.

Theorem 10-1: In a proportion, the cross-products are equal. That is, the product of the means and the product of the extremes are equal.

Translation: If $a/b = c/d$, then $ad = bc$. With real numbers, if 1:8 = 2:16, then $(1)(16) = (8)(2)$, or 16 = 16. Note that I used colons instead of fractions here. Don't worry. It's legal since both indicate a ratio.

If the product of two numbers is equal to the product of two other numbers, then the numbers can be made into a proportion. It doesn't matter which pair is designated as the means and which is designated as the extremes. The proportion remains intact.

Theorem 10-2: If the products of two pairs of numbers are equivalent, then either pair can be made the means and the other pair made the extremes in a proportion.

Translation: If $ab = cd$, then $a/b = c/d$ or $b/a = d/c$. With real numbers, if $(1)(16) = (8)(2)$ then 1:8 = 2:16 or 16:1 = 2:8.

Armed with the knowledge that cross-products of a proportion are equal, you can fill in information if it's missing. How? Get your cross-products and solve for the unknown. Here's an example:

$$\frac{x}{20} = \frac{10}{50}$$

The cross-multiplication yields the following:

$$50x = 200$$

$$x = 4$$

You can make the math a little easier on yourself by first reducing all ratios to their lowest terms. When you do so, you get numbers that are a lot easier to toss around. For example, rewrite

$$\frac{x}{20} = \frac{10}{50}$$

as

$$\frac{x}{20} = \frac{1}{5}$$

and then go from there:

$$5x = 20$$

$$x = 4$$

Fill in the x, and the complete equation reads as "4 is to 20 as 1 is to 5."

When more than one quantity is unknown, you're left with more math to do. Here's an example:

$$\frac{x+1}{6} = \frac{x+3}{9}$$

Both the first and third proportional are unknown. You may come across equations where even less may be known, but the procedure for solving the equation doesn't change. Perform some cross-multiplication, and you get

$$9x + 9 = 6x + 18$$

Now you need to solve for x. Your next step is to subtract for the left side of the equation:

$$9x + 9 - 9 = 6x + 18 - 9$$
$$9x = 6x + 9$$

Your next step is to subtract $6x$ from both sides:

$$9x - 6x = 6x - 6x + 9$$
$$3x = 9$$

Now divide each side by 3:

$$3x/3 = 9/3$$
$$x = 3$$

No one ever said that each value has to be unique. Sometimes two of the quantities in a proportation are the same. If those two quantities happen to both be the means of the proportion, then the second and third proportional are equal. In this case, the means value is referred to as the *geometric mean,* or *mean proportional,* between the first and fourth terms of the proportion. In general terms, it looks something like this:

$$\frac{a}{b} = \frac{b}{c}$$

Substituting real numbers, you get the following equation:

$$\frac{2}{4} = \frac{4}{8}$$

So 4 is the mean proportional between 2 and 8.

Proportions are very resilient. You can twist, exchange, and contort them, and they still seem to hold true. So what specifically can you do to the values of a proportion and not change the status quo?

Theorems 10-3 through 10-6 address some rules governing the transformation of a proportion into an equivalent proportion.

Theorem 10-3: The ratios on both sides of a proportion can be inverted and still remain in proportion.

Translation: If $a/b = c/d$, then it is also true that $b/a = d/c$.

Theorem 10-4: Either the means or the extremes of a proportion can change position in the equation, and the ratios still remain in proportion.

Translation: If $a/b = c/d$, then it is also true that $a/c = b/d$ or $d/b = c/a$.

Theorem 10-5: In a proportion, adding the second proportional to the first proportional and adding the fourth proportional to the third proportional results in an equivalent proportion.

Translation: If $a/b = c/d$, then it is also true that $(a + b)/b = (c + d)/d$.

Theorem 10-6: In a proportion, subtracting the second proportional from the first proportional and subtracting the fourth proportional from the third proportional results in an equivalent proportion.

Translation: If $a/b = c/d$, then it is also true that $(a - b)/b = (c - d)/d$.

Proportioning triangles

A triangle loves lines. After all, it's made up of three of them. Three to start, anyway, because you can always add more. The auxiliary lines can sometimes be more than enough to keep you busy. Speaking of auxiliary lines, one in particular comes to mind with regard to proportions — the midline. It's a line that connects two sides of a triangle at their midpoint. It is half as long as the third side of the triangle and is parallel to that side as well (see Theorems 6-3 and 6-4). Check out the triangle and its midline in Figure 10-9. I don't know about you, but I see proportions.

Figure 10-9:
The midline
of △*KLM*
creates
proportions
with the
sides it
intersects.

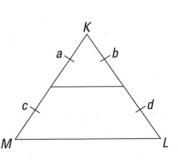

You can form three basic categories of proportions using the information in Figure 10-9's $\triangle KLM$: segment-only comparisons and two types of segment-to-side comparisons. To avoid any confusion, I want to clarify what I mean by segment and side. In this discussion, a *segment* is a portion of a side or a line — not the whole side; a *side,* on the other hand, is the whole length of any of the three sides of the triangle. Any proportion from these three categories can also be subjected to the permutations outlined in Theorems 10-3 through 10-6 that allow you to form equivalent proportions.

Proportional segments

In Figure 10-9, the two sides of $\triangle KLM$ that are intersected by the midline are split into four smaller but equal segments. The segment-only category of proportions compares the lengths of these smaller segments to each other. The proportions that are formed can be expressed in one of four directions: down, up, left, or right. The direction you use depends on how you enter the side length into the proportion.

To form a proportion of the segments formed by the midline reading in the *down* direction (see Figure 10-10), start with the segment on the upper left: a. Then move down to the second segment on the left: c. The first ratio is

$$\frac{a}{c}$$

Go to the opposite side to form the other half of the proportion. The segment on the upper right is over the segment on the lower right:

$$\frac{b}{d}$$

Put these two ratios together in a proportion, and you get

$$\frac{a}{c} = \frac{b}{d}$$

This proportion means "the upper left side of $\triangle KLM$ is to the lower left side of $\triangle KLM$ as the upper right side of $\triangle KLM$ is to the lower right side of $\triangle KLM$."

To form the proportion, you started on the left. You can start on the right if you want. Just be consistent. If you had started on the right instead of the left, the proportion would've looked like this:

$$\frac{b}{d} = \frac{a}{c}$$

Be sure that your proportion is true; the cross-products must be equal.

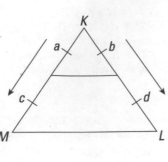

Figure 10-10:
When using
the down
direction to
form a
segments-
only
proportion,
you read
down the
sides of the
triangle.

To form a proportion reading *up,* start at the lower left (see Figure 10-11). In this example, the proportion you put together reads as "*c* is to *a* as *d* is to *b*" and looks like this:

$$\frac{c}{a} = \frac{d}{b}$$

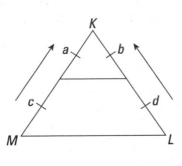

Figure 10-11:
When using
the up
direction to
form a
segments-
only
proportion,
you read *up*
the sides of
the triangle.

Reading left to right (see Figure 10-12), the proportion reads as "*a* is to *b* as *c* is to *d*" and looks like this:

$$\frac{a}{b} = \frac{c}{d}$$

Figure 10-12:
When using the left direction to form a segments-only proportion, you read from the *left to the right* of the triangle.

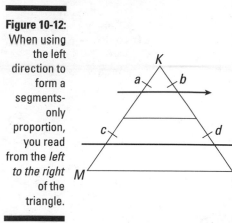

Reading right to left (see Figure 10-13), the proportion reads as "*b* is to *a* as *d* is to *c*" and looks like this:

$$\frac{b}{a} = \frac{d}{c}$$

Figure 10-13:
When using the right direction to form a segments-only proportion, you read from the *right to the left* of the triangle.

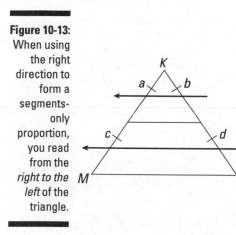

Proportional segments to sides

You can make two basic comparisons between a segment (a portion of a side) and the whole side of a triangle. You can compare the upper part of the segment to the whole, or you can compare the lower part of the segment to the whole.

When the upper segments of each side are compared to the whole side, the resulting proportion means "the upper left side of the triangle is to the whole left side as the upper right side is to the whole right side" (see Figure 10-14), and the proportion in this example is written like this:

$$\frac{a}{KM} = \frac{b}{KL}$$

Figure 10-14: Performing a segments-to-sides comparison — the upper segments to the whole sides of △KLM.

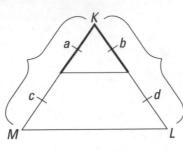

This next proportion is similar to the preceding one, with one exception: This one compares the lower segment to the whole side. This comparison reads as "the lower left segment is to the whole left side as the lower right segment is to the whole right side" (see Figure 10-15), and the proportion in this example is written like this:

$$\frac{c}{KM} = \frac{d}{KL}$$

Figure 10-15: Performing a segments-to-sides comparison — the lower segments to the whole sides of △KLM.

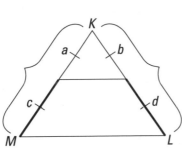

Parallel for proportions

The midline isn't the only line that forms proportions in a triangle. In fact, any line that intersects two sides of a triangle while remaining parallel to the third side divides the sides of the triangle proportionally (see Figure 10-16).

Figure 10-16:
Any line that intersects two sides of a triangle and is parallel to the third creates the opportunity for comparisons.

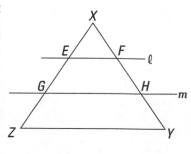

In Figure 10-16, $\overline{EF}$ and $\overline{GH}$ are parallel to $\overline{ZY}$. You've got a lot to work with here. You can compare $\overline{XE}$ and $\overline{XZ}$ to $\overline{XF}$ and $\overline{XY}$. Or you can try something new and use segments that are created by lines ℓ and m. You can compare the proportions of $\overline{EG}$ and $\overline{FH}$ to the length of their respective sides:

$$\frac{\overline{EG}}{\overline{XZ}} = \frac{\overline{FH}}{\overline{XY}}$$

Any way you slice it, all the preceding things are proportional relationships of $\triangle XYZ$ that are created by parallel lines.

Theorem 10-7: A line that is parallel to one side of a triangle divides the other two sides of the triangle proportionally.

Translation: If a line is inside a triangle and is parallel to one of the sides of the triangle, then the line divides the sides of the triangles proportionally.

Theorem 10-8: If a line divides two sides of a triangle proportionally, then it is parallel to the third side.

Translation: Theorem 10-8 just flips the information from Theorem 10-7 around, giving you a different perspective on the information. If you know that the sides of a triangle are proportionally divided, then the line that is responsible for this division is parallel to the side it does not touch.

Similar polygons

Similar polygons are like copies of each other. The thing is, unlike congruent polygons — where all corresponding parts are congruent — similar polygons have at least their corresponding angle measures in common. The sides of similar polygons aren't the same size but are proportional. Similar polygons look alike, but one of them appears to have been shrunk in the dryer (see Figure 10-17).

Figure 10-17:
Similar polygons have the same shape but not necessarily the same size.

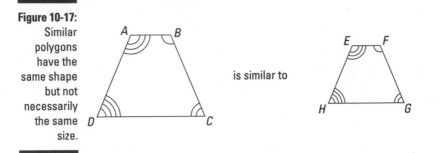

In Figure 10-17, the correspondence of angles and sides of polygons *ABCD* and *EFGH* is mapped out as follows:

Sides	*Angles*
$\overline{AB} \leftrightarrow \overline{EF}$	$\angle A \leftrightarrow \angle E$
$\overline{BC} \leftrightarrow \overline{FG}$	$\angle B \leftrightarrow \angle F$
$\overline{CD} \leftrightarrow \overline{GH}$	$\angle C \leftrightarrow \angle G$
$\overline{DA} \leftrightarrow \overline{HE}$	$\angle D \leftrightarrow \angle H$

The sides of the polygon (listed in the Sides column) are in proportion to each other, and the corresponding angles (in the Angles column) are congruent.

Two conditions must be met before you can declare two polygons similar:

✔ All pairs of corresponding angles must be congruent.

✔ All corresponding sides must be proportional.

Similar triangles

Triangles are polygons, so they don't escape the criteria required for polygons to be similar. Two triangles are similar if they meet two conditions:

All pairs of corresponding angles must be congruent, and the three pairs of corresponding sides must be in proportion to each other. Because that's the definition of similar triangles, and because all definitions are reversible, you can flip things around. Once you do that, instead of starting with the term *similar triangles* and moving on to the criteria that must be met for the two triangles to be considered similar, you start with the criteria. If the criteria are met, you have permission to declare the triangles similar. So, if you have two triangles with three pairs of congruent corresponding angles and three pairs of proportional corresponding sides, then you can declare that the two triangles are similar. (This definition reversal is useful for proving triangles similar in the proofs that follow shortly.)

Given only that the measures of the angles of similar triangles are congruent, there's no way to figure out how the sizes of the triangles compare. But you know that the sides of similar triangles are proportional. If you form a ratio of the lengths of two corresponding sides of two similar triangles, this ratio can serve as a comparison of the lengths of one triangle relative to another. Figure 10-18 shows two similar triangles. The symbol ~ is used to indicate similarity, so $\triangle ABC \sim \triangle EFG$ is translated as "$\triangle ABC$ is similar to $\triangle EFG$." $\triangle ABC$ has side lengths of 6, 8, and 10. The corresponding side lengths of $\triangle EFG$ are 3, 4, and 5.

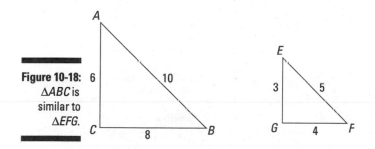

Figure 10-18: $\triangle ABC$ is similar to $\triangle EFG$.

In Figure 10-18, $\overline{AC}$ has a length of 6 and $\overline{EG}$ has a length of 3. The ratio of these two lengths is 6/3. Reduce this ratio to its lowest terms, and you get 2/1 (or 2:1). The ratio of the measure of any two corresponding sides of similar triangles (or polygons) is called the *ratio of similitude*. As you can see, this ratio tells you the size relationship of one similar triangle to another. In the case of $\triangle ABC$ and $\triangle EFG$, the ratio of similitude is 2:1, making the sides of $\triangle ABC$ twice as big as the sides of $\triangle EFG$.

The art of proving triangles similar

History does repeat itself; make no mistake about that. Although the specifics may change, the overall task and lesson remain the same. In Chapter 6, I paint a picture of how to prove triangles congruent. The process for proving triangles similar takes you down the same road. Well, not exactly the same road, but it's pretty derned close.

To prove two triangles congruent, you have to show that the three pairs of corresponding sides and angles are congruent. And you can make things easier on yourself by using some shortcuts. To prove two triangles similar, you have the daunting task of showing that the three pairs of corresponding angles are congruent and the three pairs of corresponding sides have the same ratio. Here's where history repeats itself: There are shortcuts, and they're *real* time-savers.

I'll start with the simplest shortcut. It may not be the simplest to prove, but it's the simplest to explain. If you show that any two pairs of corresponding angles are congruent, then the triangles are similar (see Figure 10-19).

Figure 10-19: If two angles of △*ABC* are congruent to two corresponding angles of △*EFG*, then the triangles are similar.

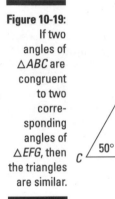

In Figure 10-19, ∠*A* ↔ ∠*E* and ∠*C* ↔ ∠*G*. It's also true that ∠*A* ≅ ∠*E* and ∠*C* ≅ ∠*G*. The figure shows that two corresponding pairs of angles are congruent, and from this info you can conclude that △*ABC* ~ △*EFG*.

Theorem 10-9: Two triangles are similar if two pairs of corresponding angles are congruent.

Translation: Two triangles are similar if they have two pairs of corresponding angles that are congruent. This one's known as the Angle-Angle Theorem of Similarity. It's abbreviated in proofs as AA.

How about a slight twist to the last example? The triangles in Figure 10-20 are both right triangles. The measure of a right angle is known. If the measure of one of the two acute angles in a right triangle is congruent to an acute angle in another right triangle, then the triangles are similar. This line of thinking follows from Theorem 10-9 and is a corollary of that theorem. Basically, you're establishing that two corresponding pairs of angles are congruent — the right angle and one of the two acute angles. In Figure 10-20, one pair of corresponding angles has a measure of 90°, and the other pair has a measure of 30°. These triangles can be declared similar.

Figure 10-20:
Two right
triangles are
similar if
one of the
acute
angles in
one triangle
is congruent
to an acute
angle in the
other
triangle.

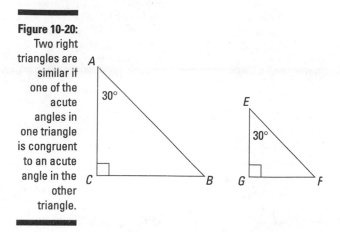

Corollary 10-1: Two right triangles are similar if an acute angle in one triangle is congruent to an acute angle in the second triangle.

In Figure 10-21, a single triangle is cut by a line that is parallel to one of the three sides while intersecting the other two. This line not only establishes the ability to make proportional statements about the segments, but it also creates a triangle that is *x* amount smaller than its host triangle.

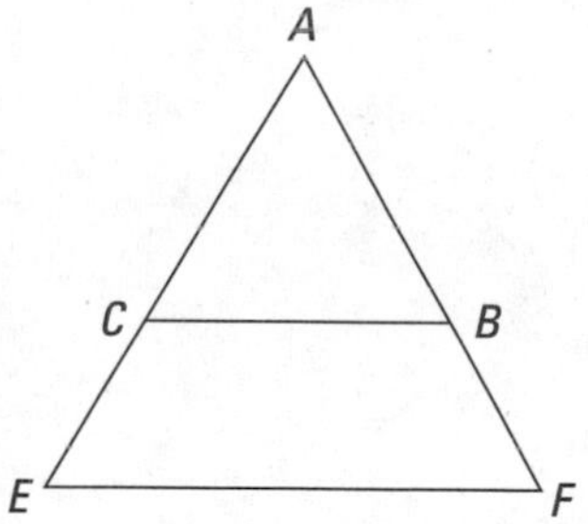

Figure 10-21:
A line
parallel to
one side of
a triangle
while
intersecting
the other
two sides
of that
triangle
creates a
smaller,
similar
triangle.

In Figure 10-21, $\triangle AFE$ shares portions of itself with $\triangle ABC$, and the measures of the three angles of these two triangles are congruent. Both triangles have $\angle A$ in common. The measure of $\angle ACB$ is congruent to the measure of $\angle E$, and the measure of ∠*ABC* is congruent to the measure of ∠*F*. How? $\overline{CB}$ is parallel to $\overline{EF}$. Now think of $\overline{AE}$ and $\overline{AF}$ as transversals of these two lines. Once you see this relationship, it's an easy leap: Corresponding congruent angles are formed when a transversal (or, here, a side of the triangle) intersects parallel lines. The sides of △*ABC* are also in proportion to the sides of △*AFE* since $\overline{CB}$ is parallel to $\overline{EF}$.

Similar triangles have their own version of the side-angle-side method of showing congruence (see Postulate 6-1) as a means of proving similarity. When trying to prove congruent triangles, you're looking for two congruent sides with an included congruent angle. Similar triangles don't require that the sides be congruent — only that the sides be proportional. That's the change. To prove that two triangles are similar, show that one pair of corresponding congruent angles is flanked by corresponding proportional sides. Sounds simple enough, right? Figure 10-22 shows what the whole shebang looks like when you've got it.

Figure 10-22:
Two triangles are similar if they have one pair of corresponding congruent angles and if the sides that make up those angles are in proportion.

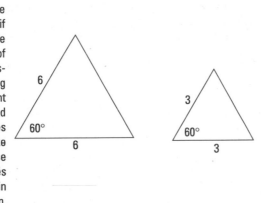

Theorem 10-10: Two triangles are similar if one pair of corresponding congruent angles has proportional lengths as the sides of the angles.

Translation: Two triangles are similar if one angle in the first triangle is congruent to a corresponding angle in the second triangle. The sides of the angles must also be in proportion to each other.

The last method for proving that two triangles are similar is also a rip-off of a method for proving congruent triangles. Similar triangles don't seem to have much originality, do they? This is good for you, though — less to remember. To prove that two triangles are congruent, you can show that the lengths of their three sides are congruent. But similar triangles aren't interested in congruency; they're interested in proportionality. Show that the three sides of one triangle are in proportion with the corresponding three sides of another triangle, and you can declare that the triangles are similar (see Figure 10-23).

Figure 10-23:
Two triangles are similar if their corresponding sides are proportional.

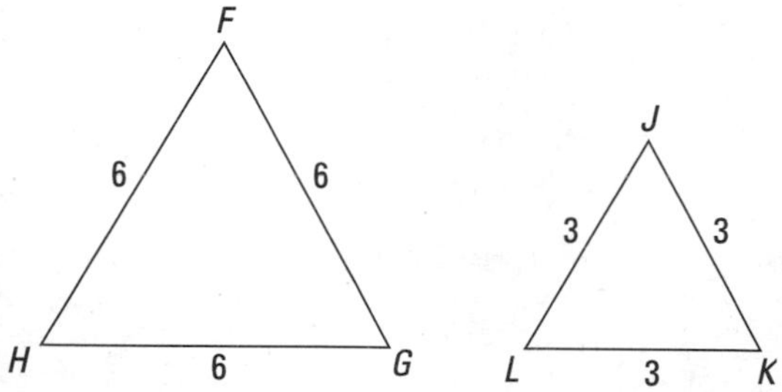

In Figure 10-23, because $\dfrac{\overline{FG}}{\overline{JK}} = \dfrac{\overline{GH}}{\overline{KL}} = \dfrac{\overline{FH}}{\overline{JL}}$ then $\triangle FGH \sim \triangle JKL$.

Theorem 10-11: Two triangles are similar if their corresponding sides are proportional.

Translation: If corresponding sides of two triangles are proportional, then the two triangles are similar. You really can't break it down any further than that.

More tidbits about proportions of similar triangles

I'd like to share some extra notes on the proportions of similar triangles. This info should give you an idea of just how far-reaching the proportionality of two triangles can extend:

✔ The perimeter of a triangle is the sum of its three sides. It follows that if the three pairs of corresponding sides of two triangles are in proportion, then the perimeters of the two triangles are also of the same proportion.

✔ If the sides of two triangles are proportional, and the perimeters of the two triangles are proportional, then the altitudes of the two triangles have the same ratio as any two corresponding sides of the similar triangles.

✔ Regardless of which method you use to prove that two triangles are similar, once you've proven that they are similar, you can cite that the corresponding sides of similar triangles are proportional (CSSTP) as a reason for proportionality of any corresponding sides in the two triangles.

Actual proofs that show similar triangles

Proof 10-1 shows that two triangles are similar if two corresponding angles are congruent.

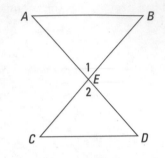

Given: $\overline{AB} \parallel \overline{CD}$.

Prove: $\triangle ABE \sim \triangle CDE$ and $\angle 1 \cong \angle 2$.

Proof 10-1:
The Triangles Similar by Angles Proof

Statements	Reasons
1) $\overline{AB} \parallel \overline{CD}$	1) Given.
2) $\angle B \cong \angle C$ and $\angle A \cong \angle D$	2) If two parallel lines are cut by a transversal, then the alternate interior angles are congruent (see Chapter 2).
3) $\triangle ABE \sim \triangle CDE$	3) Angle-Angle Theorem of Similarity (AA). If two angles of one triangle are congruent to two angles of another triangle, then the triangles are similar.
4) $\angle 1$ and $\angle 2$ are vertical angles.	4) Definition of vertical angles (see Chapter 2).
5) $\angle 1 \cong \angle 2$	5) Vertical angles are congruent (see Chapter 2).

Proof 10-2 demonstrates that if two triangles are proven similar, then any pair of corresponding sides is proportional.

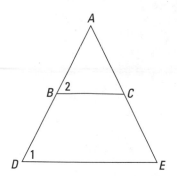

Given: $\angle 1 \cong \angle 2$.

Prove: $\dfrac{\overline{AC}}{\overline{AE}} = \dfrac{\overline{BC}}{\overline{DE}}$

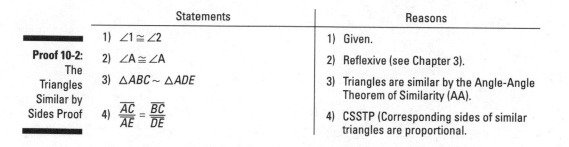

	Statements		Reasons
	1) $\angle 1 \cong \angle 2$		1) Given.
Proof 10-2: The Triangles Similar by Sides Proof	2) $\angle A \cong \angle A$		2) Reflexive (see Chapter 3).
	3) $\triangle ABC \sim \triangle ADE$		3) Triangles are similar by the Angle-Angle Theorem of Similarity (AA).
	4) $\dfrac{\overline{AC}}{\overline{AE}} = \dfrac{\overline{BC}}{\overline{DE}}$		4) CSSTP (Corresponding sides of similar triangles are proportional.

A practical application of similar triangles

You can use your knowledge of triangles to determine the height of a tall statue without measuring the statue. For example, suppose that I stand next to a tall statue. I know that I have a height of 5.5 feet and my friend, Kathy, measures my shadow to be 8.5 feet. At the same time, Debbie, measures the shadow of a statue to be 50 feet. She measures the shadow of the statue by first measuring her foot and then walking one foot in front of the other, foot tip to foot tip. How tall is the statue?

Take a look at Figure 10-24. I've laid out what I know and put an x for the value I'm seeking — the height of the statue. The resulting proportion is

$$\frac{5.5}{8.5} = \frac{x}{50}$$

Reduce the ratio on the left to its lowest terms and then cross-multiply. The resulting equation is

$$\frac{1.1}{1.7} = \frac{x}{50}$$

$$1.7x = 55$$

Solve for x by dividing each side of the equation by 1.7. You discover that the statue is approximately 32.4 feet tall.

Figure 10-24:
You can obtain the height of large objects in the real world by using a proportion.

Part V
Geometry's Odds and Ends

The 5th Wave   By Rich Tennant

You call this accurate course plotting? Now you know why they never ask you to draw crop circles anymore.

In this part . . .

Open the closet doors. *Carefully.* And duck for good measure because this closet's packed full and something might just fall on your head. Here is where all those lesser known areas of geometry are stored. You get to explore coordinate geometry, locus, trig, and solids.

(And by the way, you don't really have to duck before opening up this part because everything's organized for you in a neat and tidy way. Unlike my hallway closet. I'm *way* behind on my spring cleaning.)

Chapter 11

You Sank My Battleship! (or Working with Coordinates)

In This Chapter

▶ What coordinates and the grid are

▶ How to plot coordinates on the grid

▶ How to use the grid to find all sorts of stuff (distance, midpoints, area, slope, and the equations of a line and a circle)

*T*he hot sun beats down on the deck of the boat. Crystal blue water is laid out as smooth as a velvet carpet as far as the eye can see. The ocean lightly sways the boat as the leeward breeze blows. Then, without warning, dark clouds blanket the sky and shroud the sun. The waves that once lapped lovingly at the bow are now angrily beating a ceremonious war chant. Captain James scurries up the mast to unsnag the sail. The wrath of a raging surge jars him loose from the rigging and tosses him like a rag doll into the sea. On impact with the salt water, Captain James's transmitter on his life vest is activated: "Man overboard!" The call goes out to the Coast Guard. Captain James is fortunate to be so safety conscious. He is plucked from the sea — wet and cold but otherwise well.

That transmitter allowed the Coast Guard to locate the captain by providing his exact geographic location. An *exact* location requires two coordinates relative to an agreed-on starting point — one *latitude* (or horizontal location) and one *longitude* (or vertical location). Without them, Captain James would've been lost to the demons of the sea.

In this chapter, I introduce the grid you use to plot pairs of points that determine exact location. A grid is drawn on the globe of the earth, showing the lines of latitude that circle the earth and lines of longitude that run from the North to South Poles. Those lines allow you to find the location of any place in the world, given the proper two points. A grid is a grid no matter how you slice it, but the units may be different. So when plotting points on a grid used in geometry, you indicate location by using numerical values with positive or negative signs as opposed to the directions of north, south, east, and west. A geometric grid not only determines location but also maps out distance and area, as well as an equation for making a circle.

Describing the Grid (Sorry, It's Not Football)

The *gridiron,* as it's affectionately called, is the playing field on which a football game takes place. The movement of the football is measured in yards. The playing field is marked with yard indicators that start at midfield with the 50-yard line. They extend in either direction until they reach the end zone located at the 0-yard line at each end of the field (see Figure 11-1).

Figure 11-1:
A football playing field has distances marked off in yards from the 50-yard line in either direction to the end zone.

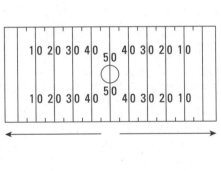

So who developed the coordinate plane?

Rene Descartes (pronounced *day-cart*) was not Greek. He was a Frenchman whose work dates back to the first half of the fifteenth century. He developed a form of analytic geometry that used a rectangular coordinate system as a means of indicating location. Using points on the plane,

Descartes developed algebraic equations to describe various geometric shapes. And so the coordinate plane is referred to as the *Cartesian plane* in fond memory of Descartes' contribution. Warms the heart, eh?

Notice that you can only determine the exact location of the ball on the football field relative to the 50-yard line and end zone. The distance of the ball from either sideline is a nonfactor, so there's no measure of it on the field. But suppose that you want to know the exact location of the ball. You can easily give the yard-line location and whether the ball is on the home- or away-team side of the field. Knowing the yard line and direction, however, is not enough to get you to the ball. You need to know, relative to the sidelines, where the ball is located. Basically, the yard-line indicators give you a horizontal location but no vertical location. Luckily, in geometry, the geometric playing field is more precise (see Figure 11-2).

Figure 11-2:
The coordinate plane — the grid you use to determine the exact location of a point.

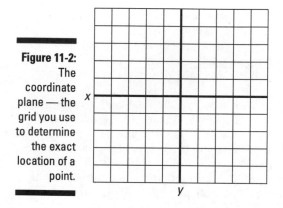

The *grid* in the geometry game is called a *coordinate plane,* and it contains a means for determining both horizontal and vertical indicators of a location. One axis is the *x,* and the other is the *y.* So which one's which? Well, the horizontal axis is the *x,* and so the vertical axis is obviously the *y* because it's the only choice remaining.

To remember which axis is which, think of the *x* as "a cross," as in *across.* The *x* goes across the paper, as in side to side, left to right. And here's a little trick to remember that the *y*-axis goes from bottom to top: Look at the letter *y.* The tail of the *y* is a line. If you don't put too much slant in the *y*'s tail, it looks kinda like the vertical axis that runs from bottom to top. I know that these tricks may seem a bit far-fetched, but they're simple and they help when you're first getting to know the material.

It Takes Two, Baby (Coordinates of a Point, That Is)

The location of a point on the grid is determined by the distance you must travel right or left along the *x*-axis and then the distance you must travel up or down the *y*-axis. The distance traveled along the *x*-axis is the *x-coordinate*. The distance for the *y*-axis is, yup, the *y-coordinate*. The *x*-coordinate is sometimes referred to as the *abscissa*, and the *y*-coordinate is sometimes called the *ordinate*. These coordinates represent the steps you must travel to get to a location on the grid — first, left or right, and then up or down. Because, to be meaningful, coordinates always travel in pairs, they're called (another easy one) *ordered pairs*. An ordered pair is made up of an *x*-coordinate and a *y*-coordinate and generically has the form (*x, y*). The *x* and *y* values represent numbers or distances to be traveled along their respective axes on the grid. Point *T*(5, 3), for example, represents a point on the grid five units to the right and three units straight up.

A quick way to properly associate the abscissa and ordinate to *x* and *y* is to think "*x* comes before *y* in the alphabet and *a* comes before *o*." The *x*-coordinate is the abscissa, and the *y*-coordinate is the ordinate. Unless, of course, you know Latin. The word *abscissa* comes from "to cut off," and the abscissa cuts off the distance to be traveled along the *x*-axis.

So how do you know where to start your travels? Simple. Every story starts at the beginning, and this one is no exception. You start from ground zero, the point at which the *x* and *y* axes intersect. Ground zero is known as the *origin*. It marks the beginning of all travel along the horizontal axis and influences the travel along the *y*-axis. Because the origin is where it all begins, it's indicated by an *x*-coordinate of 0 and a *y*-coordinate of 0. As an ordered pair, the origin is written as (0, 0) and is the point from which all other points in the coordinate system are offset.

Getting to know the quadrants

The coordinate system is divided into four quadrants. Each quadrant contains points that have locations relative to the origin. The numbering of the quadrants is I to IV, beginning at the upper right and continuing counterclockwise until you reach the lower right (see Figure 11-3).

The signs of the *x*- and *y*-coordinates determine the quadrant in which a point lies. Ordered pairs with positive *x*- and positive *y*-coordinates lie in Quadrant I. Metaphorically, Quadrant I is a happy, harmonious place. Both members of this pair have a positive outlook (☺). In Quadrant II, one member of the couple has a negative outlook (☹) while the other has a positive outlook (☺). Order pairs in Quadrant II have a negative *x* and a positive *y*. Life in Quadrant III, while harmonious, is quite grim. Both members have a negative outlook (☹). Each member of the ordered pair in Quadrant III has a negative value. And, finally, Quadrant IV has another mixed bag. Points in Quadrant IV have ordered pairs with positive *x*-coordinates (☺) and negative *y*-coordinates (☹). Check out Table 11-1 for an at-a-glance description of the four quadrants.

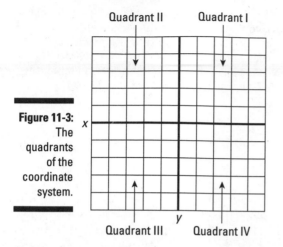

Figure 11-3:
The quadrants of the coordinate system.

Table 11-1:		The Quadrants of the Coordinate System		
Quadrant	**General Form of Ordered Pair**	**Example of Ordered Pair**	**Direction Traveled in Grid (x , y)**	**Outlook of Members in Ordered Pair**
I	(x, y)	$(5, 7)$	$(\rightarrow, \uparrow)$	(☺, ☺)
II	(x, y)	$(-3, 3)$	$(\leftarrow, \uparrow)$	(☹, ☺)
III	$(-x, -y)$	$(-6, -2)$	$(\leftarrow, \downarrow)$	(☹, ☹)
IV	$(x, -y)$	$(1, -4)$	$(\rightarrow, \downarrow)$	(☺, ☹)

Each coordinate in an ordered pair contains two important pieces of information. The positive or negative value of a number tells you the direction you need to travel to get to the location. The numeric value of the number tells you the distance (how far) you need to travel to get to the location.

Doing some plotting

Time to do some plotting. Nothing sinister, though. A point is identified on the grid by a capital letter, immediately followed by the ordered pair that identifies the location of that point. Suppose, for example, that you have a point, point A, and it has a location of $(5, 7)$ on the grid. The first thing to notice is that both numbers are positive and that the point you plot will be in Quadrant I. Next, look at the x value of the ordered pair. It's a 5. So you need to travel a distance of five units along the *x*-axis. A positive value on the *x*-axis is a step in the *right* direction. Go right from the origin five units. Next, get the *y* value. It's a 7. Because there's no negative sign, you can assume you're working with a positive 7. Locate a positive 7 on the *y*-axis and then follow the 7 line over to the right until it's directly above the *x*-axis value of 5. Mark your dot and label this point *A*(5, 7). Figure 11-4 shows *A*(5,7) in the upper right of Quadrant 1.

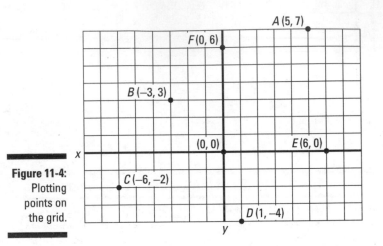

Figure 11-4:
Plotting
points on
the grid.

Now suppose that you want to plot the ordered pair (–3, 3). The *x* value of 3 has a negative sign, so move left three units from the origin. Then move up three units on the *y*-axis. Mark this point with a dot and label it *B*(–3, 3). Figure 11-4 shows this point, too.

A point that is located in Quadrant III has negative values for both *x* and *y*. The ordered pair (–6, –2) fits the bill, so this point — once plotted — will be in Quadrant III. First, plot the *x* value. It's a –6, so move six units to the left of the origin. The *y*-coordinate has a negative value, too, so it's located below the *x*-axis. Count down two units from the *x*-axis. At the given location, mark your point, label it *C*(–6, –2), and check it out in Figure 11-4.

The fourth and final quadrant has positive *x*-coordinates and negative *y*-coordinates. A point located at (1, –4) is one step to the right of the origin and four steps down. Mark this point as *D*(1, –4) and check it out in Figure 11-4.

Points with ordered pairs of (0, 6) and (6, 0) aren't located in any of the four quadrants but are on the axes of the coordinate plane. Point E is located at (6, 0) and is directly on the *x*-axis. Move six units to the right and zero units up or down. Point F is directly on the *y*-axis, with a location of (0, 6). Move zero units right or left and six units straight up. Figure 11-4 shows both of these points.

Determining Distance on the Grid

Say that you're planning a road trip to visit a friend. You whip out your map and find the town. How far away is it? You can use the scale on the map to determine how far your friend's house is from yours. Likewise, you can mark two points on the grid and use some math to determine the distance between the two points. If you want to find the distance between two points, you just have to do some subtraction to get your answer. The difference in x-coordinates gives you the horizontal distance, and the difference in y-coordinates yields a vertical distance.

In Figure 11-5, for example, point M has a location of $(-3, 5)$ and point N has a location of $(4, 5)$. If two points of a line have the same ordinate, or y-coordinate, then the distance between the two points is the length of that segment. Distance is a positive number. So, if $x_1 > x_2$ (where x_1 is the x-coordinate of the first ordered pair and x_2 is the x-coordinate of the second ordered pair), then the distance between the two points is $x_1 - x_2$. If the inequality of $x_1 > x_2$ is not true, meaning that $x_1 < x_2$ reflects the actual status of the values, then the distance between the two points is determined by subtracting x_1 from x_2, or $x_2 - x_1$. In the case of points M and N, $x_1 < x_2$, which makes the distance between them $4 - (-3)$, or 7 (see Theorem 11-1).

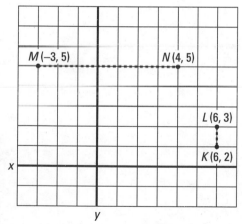

Figure 11-5:
Determining
distance on
the grid.

When plotted, two points with identical abscissas, or x-coordinates, lie on a horizontal line. The difference in the values of these two points is the length of that segment. Take point K and point L, for example. Each has an x value of 6, and they have y values of 2 and 3, respectively. If $y_1 < y_2$ (where y_1 is the y-coordinate of the first ordered pair and y_2 is the y-coordinate of the second ordered pair), the distance between the two points is $y_2 - y_1$. If, instead, $y_1 > y_2$ is true, then the distance between the two points is determined by subtracting y_2 from y_1, or $y_1 - y_2$. In the case of points K and L, $y_1 < y_2$, which makes the distance between the points $3 - 2$, or 1 (see Theorem 11-2).

Theorem 11-1: The distance between two points that have the same ordinate is the absolute value of the difference of their abscissas.

Translation: If $x_2 > x_1$, then the distance between two points with the same ordinate equals $x_2 - x_1$. If $x_1 > x_2$, then the distance between two points with the same ordinate equals $x_1 - x_2$.

Theorem 11-2: The distance between two points that have the same abscissa is the absolute value of the difference of their ordinates.

Translation: If $y_2 > y_1$, then the distance between two points with the same abscissa equals $y_2 - y_1$. If $y_1 > y_2$, then the distance between two points with the same abscissa equals $y_1 - y_2$.

What happens when you compare two points that don't have a matching x- or y-coordinate? You can probably predict it — yep, there's a formula. And here's another surprise: It's called the Distance formula. I know. Big shocker. You can get the distance between two points — (x_1, y_1) and (x_2, y_2) — by using this formula:

$$D = \sqrt{(x_2 - x_1)^2 + (y_2 - y_1)^2}$$

In Figure 11-6, for example, the distance between $T(-2, 4)$ and $W(-1, -3)$ yields the following:

$$D = \sqrt{(-1 - (-2))^2 + (-3 - 4)^2}$$
$$= \sqrt{(1)^2 + (-7)^2}$$
$$= \sqrt{1 + 49}$$
$$= \sqrt{50}$$

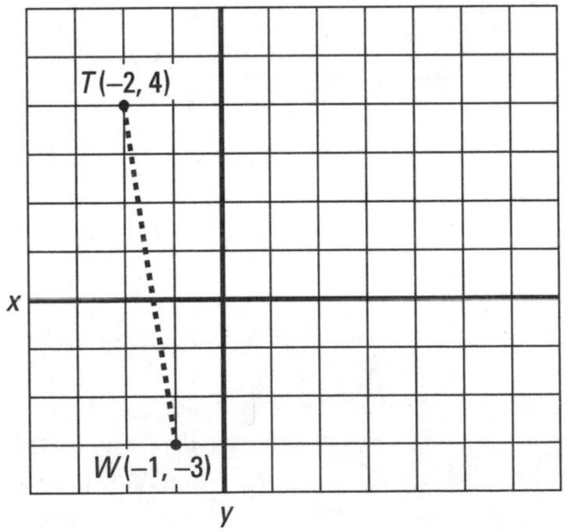

Figure 11-6: You can use the Distance formula to find the distance between point T and point W.

If a line is drawn connecting point T and point W, the distance is $\sqrt{50}$.

Theorem 11-3: The distance between two points — (x_1, y_1) and (x_2, y_2) — is determined by the formula

$$D = \sqrt{(x_2 - x_1)^2 + (y_2 - y_1)^2}$$

Translation: It's the formula. Use it.

You can apply the Distance formula to the infamous triangle. First, take stock of what you have to work with: three points connected with three lines. You can use the Distance formula to determine the distance between vertices of the triangle. After accomplishing that feat, you'll know the lengths of the sides of the triangle.

Proof 11-1, for example, uses coordinates on the grid and the Distance formula to show that a given triangle is a right triangle.

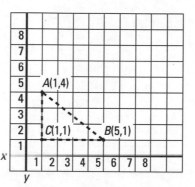

Given: $A(1,4)$, $B(5,1)$, and $C(1,1)$ are all vertices of a triangle.

Prove: ABC is a right triangle.

Before you begin: Map all the coordinates and connect the dots.

Statements	Reasons
1) $A(1,4)$, $B(5,1)$, and $C(1,1)$ are all vertices of a triangle.	1) Given.
2) $\overline{AC} = 4 - 1 = 3$	2) Subtract the ordinates to find the distance and between coordinates with matching abscissas. $\overline{AC}$ represents one side of the triangle.
3) $\overline{BC} = 5 - 1 = 4$	3) Subtract the abscissas to find the distance between coordinates with matching abscissas. $\overline{BC}$ represents one side of the triangle.

(continued)

(continued)

Statements	Reasons
4) $\overline{AB} = \sqrt{(1-5)^2 + (4-1)^2}$	4) The Distance formula for distance between two coordinates.
5) $5^2 = 4^2 + 3^2$ $25 = 16 + 9$	5) The Pythagorean Theorem, which is the formula for legs of a right triangle ($c^2 = a^2 + b^2$).
6) ABC is a right triangle.	6) In a triangle, if $c^2 = a^2 + b^2$, then it is a right $\triangle$.

Proof 11-1:
The Right
Triangle by
Coordinates
Proof

Finding a Midpoint on the Grid (or "C'mon, Meet Me Halfway")

So Trevor and Brady wanted to get together and talk about a project they'd been working on together. They lived about an hour from each other — a long distance for either one of them to travel with the limited amount of time they had. The solution? A compromise. They decided to meet each other halfway.

To determine the location that was half the distance between them, you can place the coordinates of both their locations on the grid. Represent Brady's location by point B with coordinates of (10, 10). Represent Trevor's location by point T with coordinates of (2, 1). Figure 11-7 shows the points plotted on the grid.

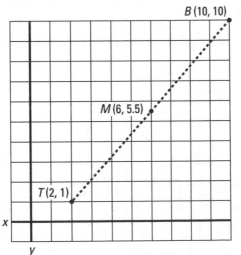

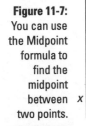

Figure 11-7:
You can use the Midpoint formula to find the midpoint between two points.

The formula to determine the distance between two points yields a single number because it's not a location but a *distance* between two points. The formula for the midpoint of two coordinates yields two values — an actual ordered pair on the line that connects the points. If you want to divide something into two equal parts, you divide it in half, right? If you have two quantities that make a whole, then you add the quantities together and then divide the sum by 2 to find the midpoint of the whole. This is the logic you need to follow in order to get the coordinates for the midpoint of a line. Follow my lead: Sum the *x*-coordinates of the two endpoints of the line and divide this sum by 2. Then do it again but this time make it the *y*-coordinates. The result is two points — one *x* and one *y* — which are the midpoint of the line. The Midpoint formula looks like this:

$$M = \frac{x_1 + x_2}{2}, \frac{y_1 + y_2}{2}$$

The two points you want to find the midpoint for are $B(10, 10)$ and $T(2, 1)$. Plug this information into the formula, and you get

$$M = \frac{10 + 2}{2}, \frac{10 + 1}{2}$$
$$= \frac{12}{2}, \frac{11}{2}$$
$$= (6, 5.5)$$

So the midpoint between the two locations is $M(6, 5.5)$.

Calculating Area on the Grid

Finding the area of a polygon using the grid is a creative process. You need to work a little more magic with the numbers before you can calculate the area. No biggie, though. This project isn't rocket science. It's made up of two types of calculations you may have already used before.

So what do you do? This:

1. **Plot your points on the grid.**

   For example, plot the four points shown in Figure 11-8: $A(2, 2)$, $B(5, 2)$, $C(5, 4)$, and $D(2, 4)$. Pretty easy so far!

2. **Draw lines connecting the dots.**

   Are you breaking a sweat yet? Just fooling. Now it's time to get down to some serious action.

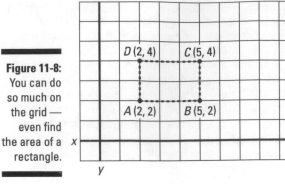

3. **Use the Distance formula to find the lengths of the sides.**

   The polygon in my example is a quadrilateral. That's four distances. Hey, I can count! Wait a second — my quad is a rectangle. I only need *two* distances because opposite sides are equal.

   $$\overline{AB} = \sqrt{(2-2)^2 + (5-2)^2} \qquad \overline{AD} = \sqrt{(2-2)^2 + (2-4)^2}$$
   $$= \sqrt{(0)^2 + (3)^2} \qquad\qquad\quad = \sqrt{(0)^2 + (-2)^2}$$
   $$= \sqrt{0+9} \qquad\qquad\qquad\quad = \sqrt{0+4}$$
   $$= \sqrt{9} \qquad\qquad\qquad\qquad = \sqrt{4}$$
   $$= 3 \qquad\qquad\qquad\qquad\quad = 2$$

4. **Locate and apply the formula for the area to your specific polygon.**

   In this case, it's the formula for a rectangle: $A = bh$ (see Chapter 7). Plug in the distances that represent the base and height, and out pops the area. (***Note:*** If you're more comfortable calling these things the length and the width, knock yourself out. The formula you use is A = *lw*. Either way you slice it, you get the same result.)

   The length, or base, of rectangle *ABCD* is 3. The height, or width, is 2. Finding the area is an easy leap from here: $A = 3 \times 2$. Alternatively, count the number of squares within the rectangle. There are 6. The area of rectangle *ABCD* is 6 square units.

How about cranking up the difficulty a notch? In Figure 11-9, I've plotted the points for another quadrilateral: *A*(2, 2), *B*(7, 2), *C*(9, 5), and *D*(4, 5). However, this time when you connect the dots, you discover that this quad is a parallelogram — not a rectangle. Don't panic. You can do this one, too. The area for a parallelogram is also $A = bh$. Just follow the steps from the preceding example, with one exception: Because calculating the area of a parallelogram requires knowing the length of its altitude (see Chapter 7), draw one. It's labeled *h* in the figure.

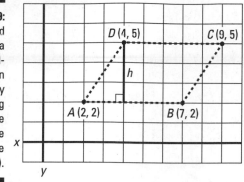

Figure 11-9: You can find the area of a parallelogram on the grid by multiplying the base by the height (the altitude).

The length, or base, of parallelogram *ABCD* is 5. The height, or altitude, is 3. $A = 5 \times 3$. The area of parallelogram *ABCD* is 15.

Now how about something just a bit different? What happens when you want to find the area of a polygon that doesn't have any horizontal or vertical sides? I . . . um . . . ah . . . subtract. Yeah, that's it — subtract. *Subtract?* Take a peek at Figure 11-10. It may look familiar if you've read Chapter 8's discussion about inscribed and circumscribed polygons. Here's what you do:

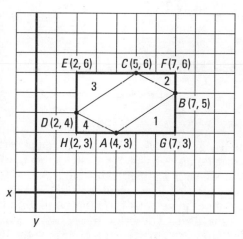

Figure 11-10: Yep, you can even use the grid to find the area of something that doesn't have any horizontal or vertical sides, like this inscribed quad.

1. **Circumscribe a rectangle about the quadrilateral.**

   Doing so wedges four right triangles between the polygon in question and the rectangle you just drew.

2. **Find the area of the rectangle by using the same steps used to find the area of the rectangle in Figure 11-8.**

   The area of rectangle *EFGH* is 15 square units.

3. **Find the area of each of the four right triangles that were created when the polygon was circumscribed by a rectangle.**

   The area of a triangle is A = ½*bh*. In Figure 11-10, for example, △*CDE* or Triangle 3 has a base of 3 units (the difference in the *x*-values from point *C* to point *E*) and a altitude of 2 (the difference in the *y*-values from point *E* to point *D*). Therefore, the area of △*CDE* equal to ½(3)(2) or 3.

4. **Sum the four areas of the right triangles.**

   In this example, the sum of the four areas of right triangles 1 through 4 is 3 + 1 + 3 + 1 = 8.

5. **Subtract the sum of the four areas of the right triangles from the area of the rectangle.**

   The area of rectangle *EFGH* is 15 square units, and the sum area of the four right triangles is 8 square units, so 15 – 8 = 7 square units. The area of the odd-shaped polygon *ABCD* is 7.

When you circumscribe a polygon about a second polygon, the sides of the circumscribed polygon are tangent to the inscribed polygon at its vertices. *Blech.* How about this, instead: When you circumscribe a polygon around another polygon, the sides of the outer polygon have to touch the inner polygon at each vertex. A little better?

Determining the Slope of a Line on the Grid

Billy and Owen's team trainer instructed them to sprint up steep hills to develop the strength of their legs. "But *steep* is such a vague term!" Owen said to me later. "How can we determine which hills would be better for our exercise program than others?" I'm a math girl, so I told them they can determine the slope of the hills by plotting a representation of them on the coordinate plane. The slope determines in numeric form the steepness of a line. (*Grade* and *pitch* are also terms used to describe steepness, or slope.) I explained to them that after they determine the steepness of the hills, they can pick the best ones for a vigorous workout.

The slope of a line (*m*) is the ratio of two quantities. You obtain the values for the ratio from two points on the line segment for which you want to determine the slope. Say you have two points, each with an x and a y value. These values will help you determine the amount of vertical change a line (Δy) has relative to the amount of its horizontal change (Δx). Cool — I just slipped in the definition of *slope*. Vertical change is represented by the y-coordinates, and horizontal change is represented by the x-coordinates. These representatives are included in the slope equation, which is

$$Slope = m = \frac{\Delta y}{\Delta x} = \frac{y_2 - y_1}{x_2 - x_1}$$

$$(x_2 \neq x_1)$$

In Formula 11-6, m represents the slope of the line. Δy is equal to $y_2 - y_1$. Δx is equal to $x_2 - x_1$.

The slope of any two points on the same line is the same.

In Figure 11-11, the slope of line $\overleftrightarrow{AB}$ is

$$Slope = m = \frac{\Delta y}{\Delta x} = \frac{y_2 - y_1}{x_2 - x_1} = \frac{6 - 2}{4 - 2} = \frac{4}{2} = \frac{2}{1}$$

In other words, for every two changes in value of the y-coordinate, there is a change of one in the x-coordinate.

Figure 11-11:
To calculate the slope of a line, divide the difference in the y values of two points by the x difference in the x values of those two points.

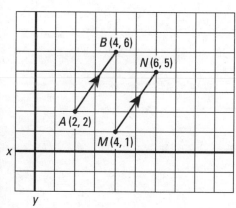

Sometimes you may see *slope* described as the ratio of rise to run. It's the same thing as the regular definition of slope — just different words. *Rise* is the vertical change, and *run* is the horizontal change.

The slope of line $\overleftrightarrow{MN}$ is equal to the slope of line $\overleftrightarrow{AB}$ (see Figure 11-11). These lines appear to be parallel, and — you know what? — they are. Lines that have the same slope are parallel. It also makes sense that parallel lines are lines with the same slope.

Theorem 11-4: The slopes of two parallel lines are equal.

Theorem 11-5: Parallel lines have equal slopes.

Translations: If lines don't intersect, it's because they have equal changes in their rise and equal changes in their run. As one line changes, the other line similarly changes in the exact direction and steepness.

The slope of a line can be positive, negative, zero, or undefined. You can usually tell the slope of a line just by looking at it. The run, or horizontal change, is the best identifier.

A line that points up to the right has a positive slope. It is a ratio of two positive numbers. Figure 11-12 shows an example. And here's the formula for that example:

$$Slope = m = \frac{\Delta y}{\Delta x} = \frac{y_2 - y_1}{x_2 - x_1} = \frac{7 - 2}{7 - 2} = \frac{5}{5} = \frac{1}{1} = 1$$

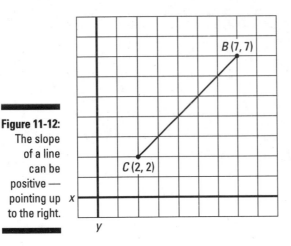

Figure 11-12:
The slope of a line can be positive — pointing up to the right.

A line that points up to the left has a negative slope (see $\overline{JK}$ in Figure 11-13). The negativeness of the slope is the fault of the negative number that results when the two y-coordinates are subtracted from each other:

$$Slope = m = \frac{\Delta y}{\Delta x} = \frac{y_2 - y_1}{x_2 - x_1} = \frac{2 - 7}{7 - 2} = \frac{-5}{5} = \frac{-1}{1} = -1$$

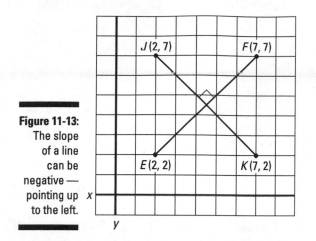

Figure 11-13:
The slope
of a line
can be
negative —
pointing up
to the left.

Read the grid left to right. A line that is higher on the grid at its left point location than at its right point location has a rise that has lost ground and is on the decline. A line with a negative slope has a rise that is falling.

In Figure 11-13, the slope of $\overline{JK}$ has a negative slope and is the reciprocal of the slope of $\overline{EF}$. $\overline{JK}$ is perpendicular to $\overline{EF}$. The slopes of $\overline{JK}$ and $\overline{EF}$ are reciprocals, and their signs are opposites:

Theorem 11-6: If two lines are perpendicular, then the product of their slopes is –1.

Theorem 11-7: If two lines are perpendicular, then the products of the slopes of the two lines is equal to –1. The slopes of those lines are negative reciprocals of each other.

Translations: Figure 11-13 does more justice than a description. A picture's worth a thousand words.

In Figure 11-14, the horizontal line has a zero slope. A line with a zero slope has no rise, or vertical change, because the y-coordinates of the two points of the line have the same y value. Any value subtracted from itself is obviously zero:

$$Slope = m = \frac{\Delta y}{\Delta x} = \frac{y_2 - y_1}{x_2 - x_1} = \frac{4 - 4}{7 - 2} = \frac{0}{5} = \frac{0}{1}$$

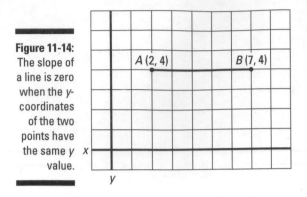

Figure 11-14:
The slope of a line is zero when the *y*-coordinates of the two points have the same *y* value.

And bringing up the rear is the vertical line. A vertical line has no slope or an undefined slope. Look at Figure 11-15. You look at the run of a line to determine its type of slope. A vertical line has no run. That is, you're comparing two coordinates with the same *x* value. As you can see from the slope equation for this line, which is

$$Slope = m = \frac{\Delta y}{\Delta x} = \frac{2-7}{5-5} = \frac{-5}{0}$$

you have to divide by zero, which is a big no-no in the math world because it yields meaningless information — thus, no slope.

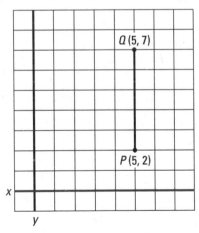

Figure 11-15:
A vertical line has an undefined slope because you can't divide by zero.

To help you remember the difference between zero slope and no slope, I can combine art and imagination: The word *zero* has a *z* in it. A *z* has two horizontal lines. Horizontal lines have a zero slope. *N* in the word *No* has two vertical lines. Vertical lines have no slope. Figure 11-16 is your visual aid.

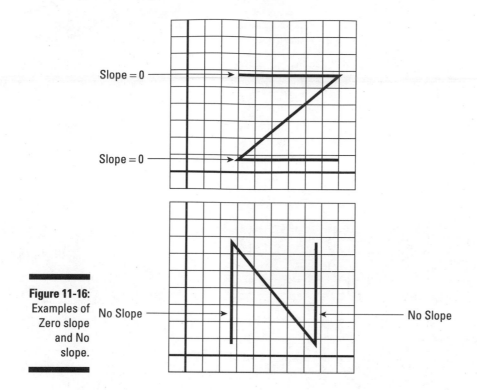

Figure 11-16: Examples of Zero slope and No slope.

All about the Equation of a Line

$$y = mx + b$$

It doesn't seem to have quite the notoriety of Einstein's $E = mc^2$. And you're probably wondering what $y = mx + b$ even is. It's the equation of a line, and each variable within the slope intercept equation of a line has a special meaning (see Figure 11-17).

Figure 11-17: The slope intercept equation of a line and what each variable represents.

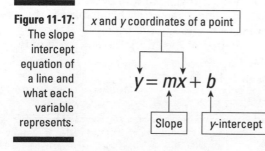

m is the slope of the line (see the section "Determining the Slope of a Line on the Grid" for what that is and how to get it). b is the *y-intercept,* which is the value of y where the line crosses the y-axis. The equation of the line may look incomplete at this point, but with an m value and a b value, the equation is ready for you to use. The x and y in the equation are actual values from an ordered pair from the grid.

In Figure 11-18, line $\overline{KP}$ has a slope of 3 and a y-intercept of -7. That info leaves you with a line equation of $y = 3x + (-7)$.

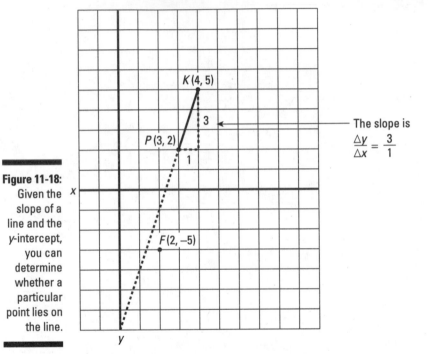

Figure 11-18:
Given the slope of a line and the y-intercept, you can determine whether a particular point lies on the line.

You can use this equation to determine whether a particular point is on this line. For example, if you have point $F(2, -5)$ and you want to determine whether point F is on your line, plug the x and y values of this point into the line equation. Substitute the x-coordinate for the x in the equation of the line and the y-coordinate for the y in the equation of the line. If the equality holds up, then point $F(2, -5)$ is a point on line $y = 3x + (-7)$:

$$-5 = (3)(2) + (-7)$$

$$-11 \neq -7$$

Because -11 is not equal to -7, point $F(2, -5)$ is not on line $y = 3x + (-7)$.

Now consider the line equation $y = 2x + 2$ and point $C(2, 6)$. Substitute the x and y values into the line equation:

$$6 = 2(2) + 2$$
$$6 = 6$$

Because $6 = 6$, point $C(2, 6)$ is on the line represented by the equation $y = 2x + 2$.

A second form for writing the equation of a line is the *point-slope form.* This form of the equation is generally more convenient to use if you already have the coordinates of two points on the line, or you know the slope of the line and know one point on the line. The point-slope line equation takes the form $y - y_1 = m(x - x_1)$. m is still the slope, and x_1 and y_1 represent x and y, respectively, of a point whose coordinates are known to be on the line. If you have a line with a slope of 2, and you know that point $C(2, 6)$ lies on this line, you can set up the point-slope form of the equation of a line by substituting in these known values (see Figure 11-19).

Figure 11-19: You can use the slope-intercept and point-slope line equation to define a line.

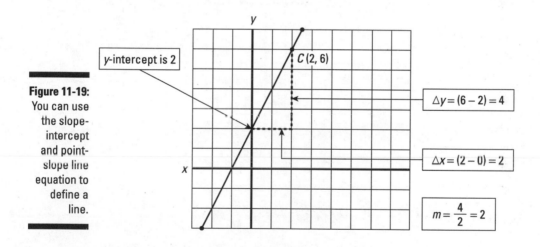

The point-slope form of the line equation is $y - 6 = 2(x - 2)$.

It doesn't matter whether you use the slope-intercept form or the point-slope form of the line equation. They both represent the same line, so they're equivalent. Don't believe me? Watch this:

$$y - 6 = 2(x - 2)$$
$$y - 6 = 2x - 4$$
$$y = 2x + 2$$

The line equation using the slope-intercept form is $y = 2x + 2$.

Lines with a zero slope and lines with no slope are special. They don't use the regular line equations discussed so far to define them. In fact, that makes life a lot simpler. A line with a slope of zero crosses the y-axis and runs parallel to the x-axis. Its line equation is given as $y = b$. b is the y-intercept. That's it for zero slope. Now for undefined slope — also known as no slope: A line with no slope is vertical, runs parallel to the y-axis, and intercepts the x-axis at a point. The equation of a line with no slope is represented by the single x-intercept value: $x = a$.

In Figure 11-20, $\overline{DL}$ is a horizontal line, so it has zero slope. The equation of the line for $\overline{DL}$ is $y = b$ or, specifically, $y = 5$. $\overline{SW}$ is a vertical line and has no slope, but it still has an equation that defines the line. The equation for line $\overline{SW}$ is $x = a$ or, as you can see from the figure, $x = 3$.

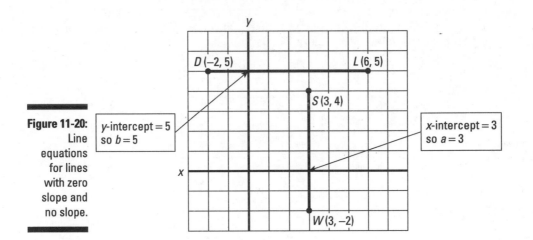

Figure 11-20:
Line equations for lines with zero slope and no slope.

All about the Equation of a Circle

Just as you can define a rectangle, parallelogram, triangle, and any other shape of polygon using the coordinate plane, you can also use the grid to find the equation of a circle. The radius of a circle is a defining piece of data. In Figure 11-21, the center of circle P is at point $P(2, 3)$, and point $J(6, 5)$ is a point on the circle. I know you can handle this next step: Draw a line from P to J. This line is the radius of the circle. If you do a small amount of cosmetic surgery on the Distance formula, you can alter the Distance formula to tailor it for the distance of the radius:

$$PJ = \sqrt{(6-2)^2 + (5-3)^2}$$
$$= \sqrt{(4)^2 + (2)^2}$$
$$= \sqrt{16 + 4}$$
$$= \sqrt{20}$$
$$\approx 4.5$$

where $\overline{PJ}$ is the length of the radius, x and y are coordinates of a point on the circle and (h, k) is the center point of the circle.

By squaring each side of the Distance formula, I can cancel out the square root sign. My Distance formula now evolves into an equation for circle P:

$$r^2 = (x - h)^2 + (y - k)^2$$

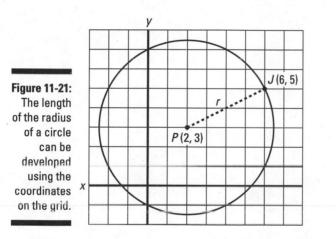

Figure 11-21:
The length
of the radius
of a circle
can be
developed
using the
coordinates
on the grid.

Theorem 11-8: A circle is defined by the equation

$$r^2 = (x - h)^2 + (y - k)^2$$

Translation: It's a formula. Fill in the values.

Note that if the center of the circle is the origin of the coordinate plane — $(0, 0)$ — the equation of the circle becomes

$$(x - 0)^2 + (y - 0)^2 = r^2$$
$$x^2 + y^2 = r^2$$

Chapter 12

Locus (and I Don't Mean Bugs)

This is a chapter on locus, *not* locust. The thought of lots of bugs really grosses me out, so I'm not even going to go there. In this chapter, I dive right in and show you how to determine and construct loci.

"If It's Not a Bug, Then What Is It?"

Good question. Glad you asked. Really. I *want* to set the record straight about exactly what a locus is. I always hate when a word has more than one meaning, especially if I'm not sure which meaning to choose. And locus is one of those words. It's obvious that a locus is not a bug or a tree because those things are each a locust. A locus can be a passage in scripture or a linear position of a gene on a chromosome. But not here. Here, a locus (singular for *loci*) still keeps its close ties to its Latin roots, meaning *place* or *location:* A *locus* is a set of points — and only those points — that satisfy a given condition or set of conditions.

From that definition, you can see that locus has that reversible thing going for it. That is, all points that satisfy a given condition(s) are included in the locus, and all points included in the locus satisfy the given condition(s). This duality is just a hazard to be endured when you're dealing with a definition. Think of a locus as the path of a point moving in accordance with the given conditions, and you won't be derailed. Enough blabbering on *that*.

Determining Loci

To determine a locus (oooh, sounds like a list), use the following steps as a guide:

1. **Make a diagram that represents all the information given.**

2. **Look for a pattern and then decide what condition(s) needs to be satisfied.**

3. **Locate a couple of other points that satisfy the condition(s).**

4. **Connect the dots.**

    A curved line or a straight line are both possible results.

    Note that this is a good opportunity to use your compass and/or ruler. I never pass up a chance to use tools.

5. **Describe the figure in the locus using a complete, grammatically correct sentence and be sure to make the conditions clear.**

That's how you determine a locus. Time to make a go of it: Given that $\overline{AB}$ is a line segment, points X, Y, and Z are all equidistant from the ends of line segment $\overline{AB}$. Make a diagram that represents this information (see Figure 12-1).

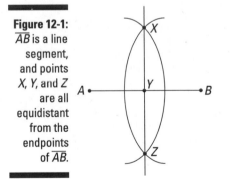

Figure 12-1:
$\overline{AB}$ is a line segment, and points X, Y, and Z are all equidistant from the endpoints of $\overline{AB}$.

See that a pattern has developed among points X, Y, and Z? They fall on a straight line. The line bisects $\overline{AB}$ and is also perpendicular to $\overline{AB}$.

The last step is to make the conditions that describe the figure.

Theorem 12-1: A locus of points equidistant from two given points is a perpendicular bisector of the line connecting those two points.

Translation: If a locus of points is equidistant from two given points, those points form a line. If you form a line segment by connecting the two given points, the line formed by the locus of points is a perpendicular bisector of that line segment.

Ready to try another one? Here's your given info: E, F, and G are points that are equidistant from the sides of angle *MNP*. Make a diagram that best represents this information (see Figure 12-2).

Figure 12-2:
E, *F*, and *G* are points that are equidistant from the sides of angle *MNP*.

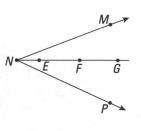

Points E, F, and G form a straight line and bisect angle *MNP*.

Theorem 12-2: A locus of points equidistant from two given intersecting lines is the bisector of any angles formed by these lines.

Translation: A locus of points equidistant from two given intersecting lines forms a line. This line is a bisector of any angles that are formed by the intersecting lines.

Here's one more example for a line, and then you can move on to loci and circles: You have a line, $\overleftrightarrow{AB}$, with points E and F on one side of $\overleftrightarrow{AB}$ and points G and H on the other side of $\overleftrightarrow{AB}$. Regardless of the side the points are located on, all points are equidistant from line $\overleftrightarrow{AB}$. See Figure 12-3. Points E and F are connected by a straight line and form $\overrightarrow{EF}$. Points G and H are connected to form $\overrightarrow{GH}$.

Figure 12-3:
Points *E*, *F*, *G*, and *H* are located on either side of $\overleftrightarrow{AB}$, but they're all equidistant from $\overleftrightarrow{AB}$.

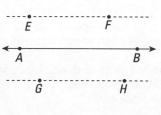

Theorem 12-3: The locus of a set of points that are an equal distance — on both sides — from a given line is a pair of lines; each of them is equidistant from and parallel to the given line, and the lines are also parallel to each other. The given line is midway between the two lines.

Translation: If you have a bunch of points located an equal distance from either side of a given line, the result is two lines that are of equal distance from the given line. The given line is exactly in the middle of the other two lines that are on either side of it.

Now here's some loci info on circles. Actually, it's *for* circles, as in making them. If you have a point P and a bunch ($n = 6$) of other points that are the same distance from point P, you have a circle (see Figure 12-4).

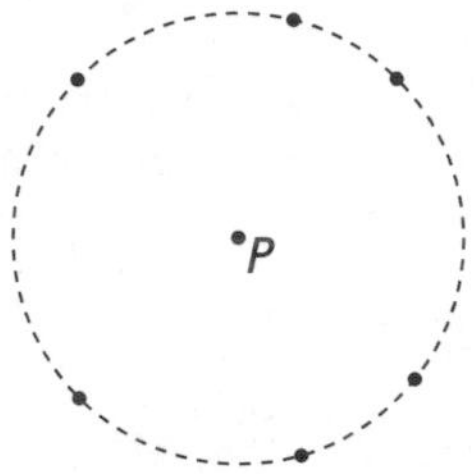

Figure 12-4:
A bunch of points that are the same distance from a given point form the locus for a circle.

Theorem 12-4: The locus of points that are a given distance from a given point is a circle whose center is the given point, and the length of the radius is the given distance.

Translation: The locus of points that are the same distance from a given point is a circle. The given point is the center of the circle, and the given distance is the radius.

Now alter the information used to create Figure 12-4. Instead of having a bunch of points an equal distance from a given point, change the info so that you have a bunch of points an equal distance from a given circle (see Figure 12-5).

Theorem 12-5: The locus of points that are a given (equal) distance from the outside of a circle is a circle outside the given circle and concentric with it.

Translation: If a bunch of points are located at an equal distance from a given circle and are located outside that circle, the points form a circle that is outside the given circle. The two circles are concentric.

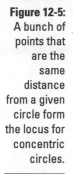

Figure 12-5:
A bunch of
points that
are the
same
distance
from a given
circle form
the locus for
concentric
circles.

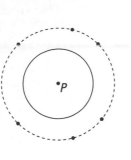

Coordinating Loci

You can add a grid when constructing loci. Adding a grid doesn't change the fundamentals — just the application of the specifics. In other words, the coordinate plane offers reference info, like the *abscissa* and the *ordinate,* that's not available in a plain plane.

"Abscissa? Oh, man." Relax. You're just experiencing coordinate geometry panic. Remain calm. Breathe. The *abscissa* is the *x*-coordinate (see Chapter 11).

A bunch of points with the same *x*-value is a vertical line, as shown in Figure 12-6. In this figure, the abscissa equals 4, and the equation of this line is $x = 4$. (For info on how to determine the value of the abscissa, turn to Chapter 11.)

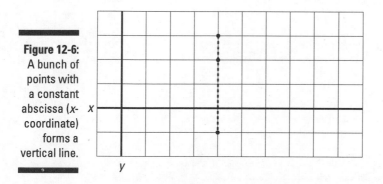

Figure 12-6:
A bunch of
points with
a constant
abscissa (*x*-
coordinate)
forms a
vertical line.

Theorem 12-6: The locus of points whose abscissa is a constant value is a vertical line parallel to the *y*-axis.

Translation: If a bunch of points have the same *x*-value, then they form a vertical line parallel to the *y*-axis.

A locus of points with the same *ordinate* (*y*-coordinate) is a horizontal line parallel to the *x*-axis. In Figure 12-7, the ordinate is equal to –2, making the equation of this line $y = -2$. (For info on how to determine the value of the ordinate, turn to Chapter 11.)

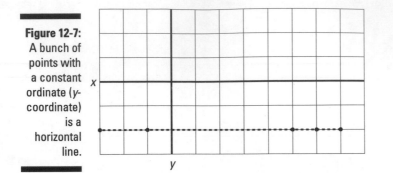

Figure 12-7: A bunch of points with a constant ordinate (*y*-coordinate) is a horizontal line.

Theorem 12-7: The locus of points whose ordinate is a constant value is a horizontal line parallel to the *x*-axis.

Translation: If a bunch of points have the same *y*-value, then they form a horizontal line parallel to the *x*-axis.

Lines, lines, lines. Try constructing a locus for a *circle* using the grid. Are you up for the challenge? First, you need a center point. Point $P(2, 3)$ will do. The locus of all points three units from point $P(2, 3)$ is a circle with a center of $P(2, 3)$ and a radius of 3 (see Figure 12-8).

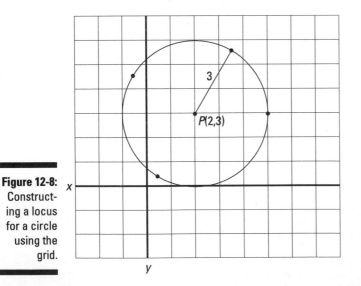

Figure 12-8: Constructing a locus for a circle using the grid.

Locating Points of Intersecting Loci

A point or points can simultaneously satisfy the conditions for more than one locus at a given time. To locate these points, you don't need a magnifying glass or a bloodhound. You just need this three-step process and a bit of brainpower:

1. **Construct a locus of the points that satisfy the first condition.**

2. **Construct a locus of the points that satisfy the second condition.**

3. **Determine the points that are contained within the intersection of the two loci.**

Say that pirates buried a treasure on a small deserted island in the Caribbean. Samantha and Abigail have found the pirates' treasure map. It indicates that the vast riches are 8 feet below the surface. Pinpointing the location to dig is a bit difficult, though. The map makes reference to something called the Devil's Gate and a two-headed serpent. From studying the geography of the island, they discover that a split-trunk coconut tree is the serpent and two towering stones are the Devil's Gate. The pirates buried the treasure five paces from the coconut tree and equal paces from each of the stone towers. Take a look at Figure 12-9. Locate all the points that are five paces from the coconut tree. These points form a circle around the tree. The circle satisfies the condition for one locus. For the second locus, locate all points that are equal paces from both of the stone towers. This locus of points forms a line that is perpendicular to the line that connects the two stone towers. Determine the point that the two loci have in common. Two points satisfy the two loci described in the map — X and Y.

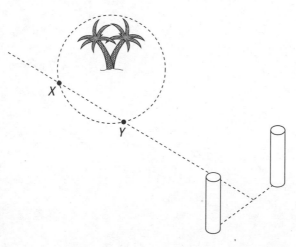

Figure 12-9:
This treasure map shows an example of inter-secting loci.

Proving Loci

After you're exposed to loci theorems, you don't need to get vaccinated. But you do need to get verification. One way to prove that a locus theorem is correct is to prove that *both* of the following statements are true:

- ✔ All points that satisfy a given condition(s) are included in the locus.
- ✔ All points included in the locus satisfy the given condition(s).

This isn't the only path to finding truth, however. You can also prove a locus true by showing *both* that

- ✔ A point that is on the locus satisfies all the given conditions.
- ✔ A point not on the locus does not satisfy the given conditions.

The method you use will, more likely than not, be dictated by the given information.

Now how about a locus proof for Theorem 12-1? I'll just restate it here so that you don't have to flip through the chapter looking for it: A locus of points equidistant from two given points is a perpendicular bisector of the line connecting those two points. Whichever method you use, proving a locus requires that both conditions be met, so you need two proofs — one for each condition. Proof 12-1 demonstrates that under the given conditions, the following locus is true: If a point (X) is equidistant from two points (M and N), then it lies on the perpendicular bisector ($\overline{XY}$) of the line segment ($\overline{MN}$).

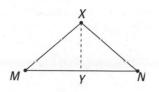

Given: $\overline{MN}$ is a line segment, and X is a point not on $\overline{MN}$, and $\overline{MX} \cong \overline{NX}$.

Prove: X is a point that lies on the perpendicular bisector of $\overline{MN}$.

Before you begin: Draw $\overline{XY}$ from point X so that it bisects $\overline{MN}$.

Statements	Reasons
1) $\overline{MN}$ is a line segment, and X is a point not on $\overline{MN}$.	1) Given.
2) $\overline{MX} \cong \overline{NX}$	2) Given. *(side)*
3) $\overline{MY} \cong \overline{NY}$	3) Y is the midpoint of MN (drawn). *(side)*
4) $\overline{XY} \cong \overline{XY}$	4) Reflexive (see Chapter 3). *(side)*
5) $\triangle MYX \cong \triangle NYX$	5) SSS. Triangles are congruent by side-side-side (see Chapter 6).
6) $\angle XYM \cong \angle XYN$	6) By CPCTC. (Corresponding parts of congruent triangles are congruent). $\angle XYM$ and $\angle XYN$ are congruent right angles that lie on line MN (see Chapter 6).
7) $\overline{MN} \perp \overline{XY}$	7) Lines that meet at a right angle are, by definition, perpendicular.
8) $\overline{XY}$ is a perpendicular bisector of $\overline{MN}$.	8) By definition (because $\overline{XY}$ is drawn as a perpendicular bisector of $\overline{MN}$).
9) X is a point that lies on $\overline{XY}$.	9) By definition (because X is an endpoint of $\overline{XY}$).
10) X is a point that lies on the perpendicular bisector of $\overline{MN}$.	10) Lines that meet at a right angle are, by definition, perpendicular.

Proof 12-1:
The Locus for Perpendicular Bisector Proof: Part 1

Proof 12-2 demonstrates that point X is on a perpendicular bisector of $\overline{MN}$ and that point X is equidistant from point M and point N.

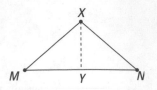

Given: Point X is a point on $\overline{XY}$, and $\overline{XY}$ is a perpendicular bisector of $\overline{MN}$.

Prove: X is equidistant from M and N.

Proof 12-2:
The Locus
for Perpen-
dicular
Bisector
Proof:
Part 2

Statements	Reasons
1) Point X is a point on $\overline{XY}$.	1) Given.
2) $\overline{XY}$ is a perpendicular bisector of $\overline{MN}$.	2) Given.
3) $\angle XYM \cong \angle XYN$	3) A perpendicular bisector forms right angles (see Chapter 2). *(side)*
4) $\overline{XY} \cong \overline{XY}$	4) Reflexive (see Chapter 3). *(angle)*
5) $\overline{MY} \cong \overline{NY}$	5) By definition, a perpendicular bisector ($\overline{XY}$) forms two congruent segments at the point of intersection with $\overline{MN}$. *(side)*
6) $\triangle MYX \cong \triangle NYX$	6) SAS. Triangles are congruent by side-angle-side (see Chapter 6).
7) $\overline{XM} \cong \overline{XN}$	7) CPCTC (see Chapter 6)
8) X is equidistant from M and N.	8) $\overline{XM} \cong \overline{XN}$

Chapter 13
Trigonometry Rears Its Ugly Head

* *

* *

*T*rigonometry is a big word. I don't know about you, but I've always hated big words. They sound sooo intellectual. Anything over three syllables gets on my nerves, but I guess the five syllables of *trigononmetry* are necessary to get in the full meaning of the word. Trigonometry comes from the Greek word *trigonometria,* meaning "triangle measure." If I were to stop at just three syllables, I'd only be able to get in the triangle part, so I guess that's my proof that big words are sometimes necessary. Nice thing is, this big word is sometimes abbreviated as *trig*. Short and sweet. I like that better and use the abbreviation in this chapter.

The properties of right triangles are the basis for trig. In this chapter, you discover methods for finding measurements of angles and sides of right triangles from the measures of parts of the triangles that are known. To get these measurements, you have to work a bit of trigonometric magic because the measurements aren't always practical or possible to get using conventional means — the ruler.

Right Triangles: A Quickie Brain Dump

A right triangle has three sides and three angles. One angle is obviously a right angle. The other two angles are acute. The side opposite the right angle is the hypotenuse. The legs are the other two sides. Whether a side is referred to as opposite or adjacent depends on the angle you use as your reference. A side that is opposite of one acute angle is adjacent to the other acute angle. A given leg can't be both opposite of and adjacent to the same angle at the same time. But, if you switch the angle from which you're viewing the sides, you switch the label of the sides so that they appropriately correspond to the angle you're working with. Figure 13-1 shows the various parts of a right triangle.

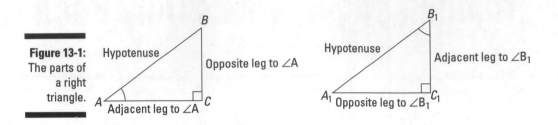

Figure 13-1:
The parts of a right triangle.

Right triangles and similarity

Think of the right triangles in Figure 13-1 as two identical wedges of cheese, which means that $\triangle ABC \cong \triangle A_1 B_1 C_1$. Now say that I've got a party starting in two hours, which is what the cheese is for, but I'm getting pretty hungry. So I take a knife and cut about an inch off the back side of one wedge of cheese (see Figure 13-2). Now my two cheese wedges are no longer congruent; they're just similar.

Figure 13-2:
Cutting one triangle makes the two triangles from Figure 13-1 similar but no longer congruent.

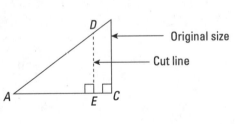

Original size

Cut line

Similar triangles have the same shape but not the same size. They're in proportion to each other. Scaled-down drawings use proportional reductions in size.

If two right triangles are similar to each other, the ratios created from corresponding parts are also similar (see Chapter 10). The right triangles formed by my cheese wedges are similar by the Angle-Angle Theorem of Similarity (see Theorem 10-9). Figure 13-2 shows an overlay of the wedge I sampled over the wedge I didn't touch. Because the wedges both started as the same size, and I made a cut off the back of one, the pointy angle is still the same measure for both of them. I made a straight cut off the back and preserved the right angle so that the two wedges also have the measure of the right angle in common. So the two triangles aren't the same size but have the measures of two angles in common, which means they're similar.

I can now go crazy. Once I've proven that two triangles are similar, I can use the argument "corresponding parts of similar triangles are proportional" to show whatever I want (within reason, of course). Which means

$$\frac{\overline{DE}}{\overline{AD}} = \frac{\overline{BC}}{\overline{AB}} = \frac{\text{Length of leg opposite } \angle A}{\text{Length of hypotenuse}}$$

I'm sure you're sitting there thinking, "So?" Well, this similarity is the basis on which you can use measurements that you *can* obtain to *get* measurements you can't. This means, revelation time, that the side-length ratios of any acute angle of a given right triangle are the same regardless of the size of the right triangle. This is how the math gods can make those wonderful trig tables (see Appendix A).

Right triangles, ratios, and two special angles

When you're working with right triangles, several ratios are available for finding the distances and angle measures of things that aren't practical to measure. These ratios include sine, cosine, and tangent — the Three Musketeers in this arena. Other ratios are available, too. These ratios are discussed in this chapter.

Which ratio you choose to use depends on what you know and what you don't. You can't use a ratio if you don't have all its required parts. You must also be sure that your information forms a right triangle.

With that info in mind, you can use two general types of angles to gain access to the immeasurable: angles of elevation and angles of depression. Each gets its name from its view of the situation relative to a horizontal line. *Angles of elevation* open upward, and *angles of depression* are downers (see Figures 13-3 and 13-4).

In Figure 13-3, the captain of a boat uses a scope to sight the top of a lighthouse. The angle he tilts his scope upward from his horizontal line of sight so that he can see the top of the lighthouse is the angle of elevation.

In Figure 13-4, just as the captain can look up to the lighthouse, the light keeper in the lighthouse can look down to the boat. The light keeper tilts his scope downward from its horizontal position to see the boat; the angle that he tilts his scope downward is the angle of depression. The light keeper's original horizontal line of sight is parallel to the ocean on which the boat floats. This means that the angle of depression is of the same measure as angle B (at the boat) because alternate interior angles formed by parallel lines are congruent.

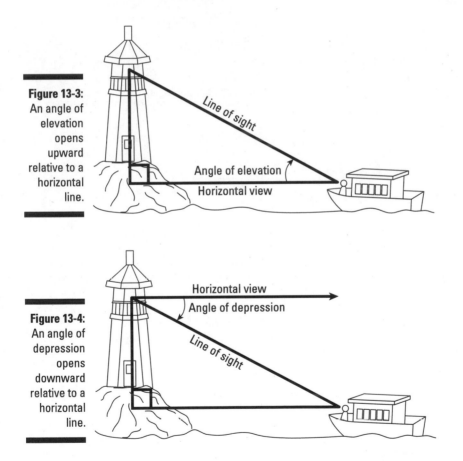

Figure 13-3:
An angle of elevation opens upward relative to a horizontal line.

Line of sight

Angle of elevation

Horizontal view

Figure 13-4:
An angle of depression opens downward relative to a horizontal line.

Horizontal view

Angle of depression

Line of sight

Give Me a Sine

The comparison outlined in the ratio

$$\frac{\text{Length of leg opposite } \angle A}{\text{Length of hypotenuse}}$$

is known as the *sine ratio*. This ratio remains the same if the sides are enlarged or reduced proportionately. But the ratio has a direct relationship to the measure of the angle. The sine ratio becomes smaller as the measure of the angle from which this measurement is taken gets smaller. After all, it *is* a comparison.

- Sine is sometimes abbreviated as *sin*. To remember that the sine of an angle is the opposite side length over the hypotenuse length — I don't mean this in a preachy way — try thinking that *sin* is the opposite of something you want to do. Just remember that in trig it's pronounced like the word "sign."

- The Greek letters α, β, φ, γ, and θ — alpha, beta, phi, gamma, and theta — may be used to represent the measure of an angle.

- Sines and cosines have values ranging from 0 to 1. An angle with a degree measure of 0° has a sin θ value of 0.0000 and is written as sin 0° = 0.0000. An angle with a degree measure of 90° has a sin θ value of 1.0000 and is written as sin 90° = 1.0000. The cosine's values run in the opposite direction: cos 90° = 0.0000 and cos 0° = 1.0000. (Check out the section "The Scoop on Cosine in Relation to Sine" for details on cosine. Its abbreviation, by the way, is *cos.*)

Sine in action: Some examples

Look at Figure 13-5 for a couple of sine examples. The sine of angle *A* is equal to the length of the side opposite angle *A* over the length of the hypotenuse, or *a/c*. The sine of angle *B* is equal to the length of the side opposite angle *B* over the length of the hypotenuse, or *b/c*.

Figure 13-5:
The sine of angle *A* is the ratio of the side opposite angle *A* over the hypotenuse.

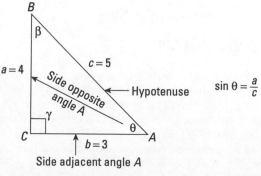

$$\sin \theta = \frac{a}{c}$$

In Figure 13-5, the lengths of sides *a*, *b*, and *c* of right △*ABC* are 4, 3, and 5, respectively. The sine of angle *A* can be written as

$$\sin A = \sin \theta = \frac{a}{c} = \frac{4}{5} = 0.8$$

And the sine of angle *B* can be written as

$$\sin B = \sin \beta = \frac{b}{c} = \frac{3}{5} = 0.6$$

To find the measures of angle *A* and angle *B*, you can use the decimal you just got from the sine ratio. For angle *A*, or sin θ, the value is sin 0.8. You can do one of two things with this number. What you do depends on your resources. That is, do you have a scientific calculator or a trig table? Either way, you should get familiar with how to use both resources so that you have the option to choose.

Finding sine measures with a trig table

With the trig table option and the info from Figure 13-5, you can look up the decimal in the sin θ column that's closest to your decimal value of sin 0.8. (Appendix A contains a trig table, if you don't have one.) It's unlikely that you're going to find exactly what you're looking for every time. Reading down the column, find the decimal that's closest in value to 0.8. When I look, for example, I come across 0.7986 and 0.8090. The value of 0.8 is in between these two values. So, next, you need to get the difference between your decimal and the ones in the table: 0.8090 – 0.8 = 0.0090. 0.8 – 0.7986 = .0014. That makes 0.8 a little closer to 0.7986. Next, adopt its corresponding angle measure in the table, which is 53°. So the degree measure of angle *A* is 53°. For angle *B*, you have a decimal value of sin 0.6. Reading down the column, you see that the closest decimal is 0.6018. According to the table, angle *B* (or β) has a measure of 37°.

Note: Although this example utilizes the trig table to get info for sin θ, it can also be used to get info on cos θ, and tan θ. I will get to those shortly.

Finding sine measures with a scientific calculator

If you decide to take the calculator route, you need a scientific calculator that has trig functions on it. The run-of-the-mill plus and minus signs aren't much help. In Figure 13-5, sin *A* equals 0.8. The buttons you press on your calculator to get the degree measure of angle *A* depends on the calculator you have. Generally, there's an INV function (for inverse) or 2^(nd) F or Shift key. Locate the key your calculator has because you're gonna need it. Punch in the number 0.8 and then press the INV or 2^(nd) F or Shift key. Press the sin key. Round the number in the display to the nearest degree. In this case, sin A of 0.8 equals 53.130102. Rounded, it's 53°. You can use the same procedure for sin B.

Note: This calculator example uses sin θ info. However, the calculator can also be used with info for cos θ, and tan θ. I will get to those shortly.

The Scoop on Cosine in Relation to Sine

In the preceding section, you find the sine of angle A and the sine of angle B in an example. But you can get the sine of angle B in a different way — by getting the sine of angle A and the *cosine* of angle A. The cosine of angle A is equal to

$$\cos A = \cos \theta = \frac{b}{c} = \frac{3}{5}$$

and is equivalent to sin B. In a right triangle, the cosine ratio of an acute angle is the ratio of the length of the adjacent side over the hypotenuse.

You can form two sine ratios and two cosine ratios using the acute angles in a given right triangle:

$$\sin A = \sin \theta = \frac{a}{c}$$
$$\cos A = \cos \theta = \frac{b}{c}$$
$$\sin B = \sin \beta = \frac{b}{c}$$
$$\cos B = \cos \beta = \frac{a}{c}$$

In a right triangle, the relationship between sin θ and cos θ stems from the Pythagorean Theorem. (The Pythagorean Theorem is $a^2 + b^2 = c^2$.) The relationship is $sin^2\,\theta + cos^2\,\theta = 1$, where θ (pronounced *theta*) is the measure of an acute angle of the right triangle.

$Sin^2\,\theta$ is equal to $(\frac{a^2}{c})$ and $cos^2\,\theta$ is equal to $(\frac{b^2}{c})$. Taken together, $\frac{a^2}{c} + \frac{b^2}{c} = \frac{a^2 + b^2}{c^2}$ You know from the Pythagorean Theorem that $a^2 + b^2 = c^2$. So substitute c^2 for $a^2 + b^2$. The result is $sin^2\,\theta + cos^2\,\theta = \frac{c^2}{c^2}$ which equals 1.

Tangents, Cotangents, Secants, and Cosecants

Sine and cosine are only two of several types of trigonometric ratios that you can form using the lengths of the sides of a right triangle. So what else do you have? This section sorts out the other ratios.

The tangent ratio

In a right triangle, the tangent ratio for an acute angle is

$$\tan A = \tan \alpha = \frac{a}{b} = \frac{\text{Length of side opposite } \angle A}{\text{Length of side adjacent } \angle A}$$

The tangent ratio is opposite over adjacent (see Figure 13-6). I remember having this info *drilled* into my head. Try to get it drilled into yours, too.

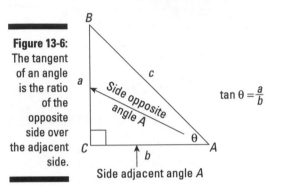

Figure 13-6: The tangent of an angle is the ratio of the opposite side over the adjacent side.

Note: Although the tangent ratio is opposite side length over adjacent side length, you may sometimes see it written as

$$\tan \theta = \frac{\sin \theta}{\cos \theta}$$
$$\theta \neq 90°$$

It's the same thing. Take a look:

$$\frac{\sin \theta}{\cos \theta} = \frac{\frac{a}{c}}{\frac{b}{c}} = \frac{a}{c} \div \frac{b}{c} = \frac{a}{c} \times \frac{c}{b} = \frac{a}{b} = \tan \theta$$

Like the sine ratio, the tangent ratio is influenced by the measure of the acute angle that dictates the ratio. But unlike the sine and cosine ratios, the value of the tangent is not constrained. It's free to roam. Its values start at zero and continue onward and upward from there. Tan 0° equals 0, marking the minimal tangent value. By the time you hit 90 degrees, tan 90° has an immeasurable, or undefined, value. (*Tan* is the abbreviation for tangent, by the way.)

Some tangent examples

Although you can use the following steps with any trig ratio, they are more easily applied to real-world examples using tangents. Why? Because of the ease in getting the unknown info.

1. **Make a scaled drawing that contains all the known information for the line segments and angles.**

2. **Mark with a variable any information that needs to be obtained.**

   You can use whatever you want as long as it doesn't confuse you. For example, use θ (theta) for a missing angle measure and x for a missing segment length.

   You can use the tangent ratio to fill in the blanks if you have a right triangle in which you know the measures of one leg and one acute angle, and you want the length of the other leg.

3. **Assess the drawing and determine what information can be used to get the info you want.**

4. **Write out the equation and solve it.**

Now how about some fun? Try these tangent examples:

Find the height of a tree that is 30 feet from you. The angle of elevation to the top of the tree is 35° (see Figure 13-7).

Figure 13-7:
You can find the height of a tree by using the angle of elevation and the distance it is sited from the tree.

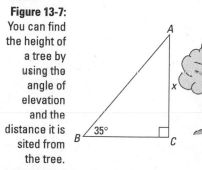

Look at the information you have and the information you need. You have an acute angle measure and the adjacent side. You want the opposite side.

$$\tan A = \frac{\text{Length of side opposite } \angle A}{\text{Length of side adjacent } \angle A}$$

$$\tan 35° = \frac{x}{30}$$

$$0.7002 = \frac{x}{30}$$

$$x = 30\,(0.7002)$$

$$x = 21.006 \text{ feet}$$

Notice in the formula that tan 35° is 0.7002. You can get this info from a trig table or a scientific calculator (see the sections "Finding sine measures with a trig table" and "Finding sine measures with a scientific calculator" for the details, which work with tangents too). Using the table, look up tan θ for θ of 35°. Using the calculator, type in 35 and press the tan button.

When all is said and done, the tree comes out to be approximately 21 feet tall.

Suppose that your dad fights fires (as mine does) with the National Forestry Service, this one's for him: Although aerial spotters are generally used now, this example goes back to the old days of lookout towers. A lookout tower is 250 feet tall. A spotter in the tower identifies a fire burning in some brush off in the distance. The angle of depression is 10°. How far away from the tower is the fire? Figure 13-8 is your visual aid. Note that angle F is the fire, angle T is the top of the tower, and angle C is a right angle.

Information evaluation time. You have an acute angle and an opposite side. Also, angle F has the same measure as the angle of depression because alternate interior angles of parallel lines are congruent.

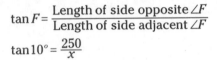

$$\tan F = \frac{\text{Length of side opposite } \angle F}{\text{Length of side adjacent } \angle F}$$

$$\tan 10° = \frac{250}{x}$$

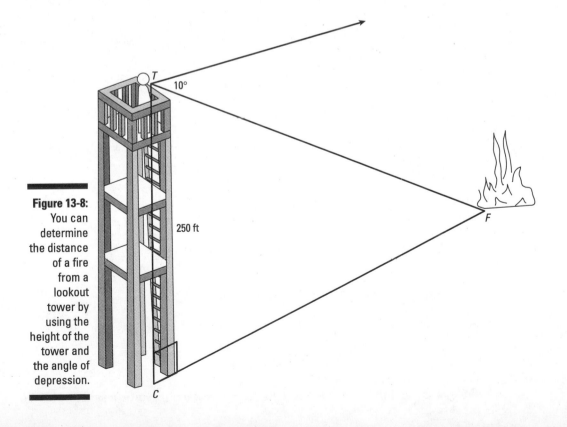

Figure 13-8:
You can determine the distance of a fire from a lookout tower by using the height of the tower and the angle of depression.

Before you can proceed any further, you must get tan 10°. You can go the trig table route or the scientific calculator route just like when you're working with sin θ (see the sections "Finding sine measures with a trig table" and "Finding sine measures with a scientific calculator" for details). Using the table, look up tan θ for θ of 10°. Using the calculator, type in 10 and press the tan button.

$$0.1763x = 250$$
$$x = \frac{250}{0.1763}$$
$$x = 1418.0374 \, \text{feet}$$

The fire is located approximately 1,418 feet from the lookout tower.

The other ratios: cotangent, secant, cosecant

The cotangent of an angle is the inverse of its tangent. That makes it the adjacent side length over the opposite side length:

$$\cot \theta = \frac{1}{\tan \theta}$$
$$\tan \theta \neq 0°$$

Cot θ is also known as the reciprocal of tan θ.

Secant θ is the reciprocal of cos θ when cos θ ≠ 0. When you boil it down to the actual sides, it's the hypotenuse over the length of the adjacent side:

$$\sec \theta = \frac{1}{\cos \theta}$$
$$\cos \theta \neq 0$$
$$= \frac{1}{\cos \theta} = \frac{1}{\frac{b}{c}} = \frac{c}{b}$$

Cosecant θ is the reciprocal of sin θ when sin θ ≠ 0. The sides involved in this ratio are the hypotenuse over the length of the opposite side:

$$\csc \theta = \frac{1}{\sin \theta}$$
$$\sin \theta \neq 0°$$
$$= \frac{1}{\sin \theta} = \frac{1}{\frac{a}{c}} = \frac{c}{a}$$

Chapter 14

The Third Dimension: Looking at Solid Geometry

• •

In This Chapter

▶ Understanding what makes an object 3-D

▶ Looking at various 3-D objects (prisms, pyramids, cylinders, cones, and spheres)

• •

*H*ave polygons left you feeling flat? Well, that can happen when you're dealing with two-dimensionals; they only have length and width. No need to fear, though. 3-D is to the rescue, and you don't even have to wear those funky glasses. In the third dimension, figures actually have some depth.

In this chapter, you find out what life in the third dimension actually means. And don't worry. You don't have to ponder the meaning of life. Just more information and, of course, more formulae.

Polywhatta? Polyhedra (or Polyhedrons, If You Prefer)

A polygon is a two-dimensional, multiangled figure. All the points of a polygon lie on the same plane, which accounts for its flatness. *Bor-ing.* In the 3-D world, the counterpart to a polygon is a polyhedron. In its Greek origins, a polyhedron is touted as being multisided. Its points exist across intersecting planes, which gives it depth and makes it solid.

Polyhedra (or *polyhedrons,* as some people refer to them) are solid geometric shapes bounded by multiple planes. When the points of a geometric shape lie on more than one plane, the shape gets some depth. This is true of the polyhedron, but the number of planes involved depends on how many sides the polyhedron has. Each side of the polyhedron forms an edge of the figure. The edges of a polyhedron are formed by the intersection of two planes that create angles known as *dihedral angles*. When the adjacent sides of the polyhedron are affixed together to form congruent dihedral angles, then the result is a regular polyhedron. Figure 14-1 gives you a glimpse of some regular polyhedra. Notice that they're named by the number of their sides.

Figure 14-1:
Polyhedrons
are solid
geometric
shapes with
sides on
multiple
planes.

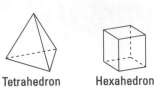

Tetrahedron   Hexahedron

Prisms: Polygons with Depth and Personality

A crystal prism dangling in a sunny window reflects a rainbow of colors on the wall. That's how I remember prisms: Prisms refract light and display the colors of the visual spectrum. That's enough with the science lesson. I don't want to distract you with the pretty colors.

Prism characteristics

A *prism* is a polygon that has been filled with space. To understand what I mean, take a triangle and add solid space to it: Draw two congruent triangles — each on a separate piece of paper — and then smoosh the papers together, one exactly on top of the other. The two pieces of paper represent separate but equal parallel planes. As you pull the pieces of papers apart, the triangles stretch, forming a solid figure. It's like stepping on gum and having it stick to your shoe. Figure 14-2 shows two triangular prisms. They are called triangular prisms because their base polygons are triangles. If the base polygons were rectangles, then the prisms would be rectangular prisms. Is this cool or what?

Figure 14-2:
A prism is a polygon that's been filled with space; its bases are on separate, parallel planes.

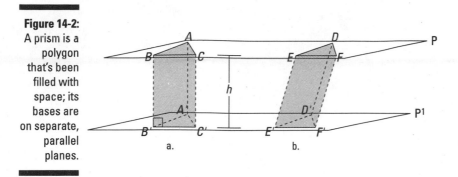

a.   b.

The farther you pull the bases of a prism apart from each other, the longer the height, or altitude, becomes. In Figure 14-2, the two congruent triangles on the separate, parallel planes (or pieces of paper) are the *bases* of the prism. The base on plane P is the upper base, and the base on P^1 is the lower base. The sides of a base are called *base edges*. In Figure a of Figure 14-2, for example, $\overline{AB}$, $\overline{BC}$, $\overline{AC}$, $\overline{A'B'}$, $\overline{B'C'}$, and $\overline{A'C'}$ are base edges. The other sides that aren't bases are called *lateral faces* or *lateral sides*. Lateral faces are bounded by lateral edges.

Prism personality profiles

Two general types of prisms exist. A *right prism* is shown in Figure a of Figure 14-2, and an *oblique prism* is shown in Figure b. You can look at a right prism in two ways: A right prism has lateral edges that meet perpendicularly with the base edges, or the altitude perpendicularly meets the plane. Oblique prisms, on the other hand, are a bit slanted.

Prism areas

You've got choices here. Do you want the lateral area or the total area?

The lateral area is the sum of all the areas of the lateral faces of the prism. Lateral faces can have a rectangular shape (for right prisms) or parallelo-gram-ular shape (for oblique prisms), so the area for each lateral side is base times height. In the case of a prism, the value for base is the length of the base edges of the polygon on the plane. The sum of the lengths of the base edges is the perimeter of the base. The total area also throws the areas of the bases into the mix. The total area can also be referred to as the total surface area. Because you may need to calculate the total area someday, get a feel for the procedure by looking at Figure 14-3 and the following formulae as applied to the various parts of the figure.

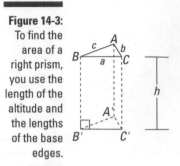

Here's the lateral area formula for the right prism in Figure 14-3:

$$S = ah + bh + ch$$

$$S = h(a + b + c)$$

$$S = hP$$

where $P = a + b + c$ or the perimeter of the base of the prism. This formula is good for any right prism.

Here's the total area formula for the right prism Figure 14-3:

$$T = S + 2B$$

where $B = \sqrt{S(S - a)(S - b)(S - c)}$ and where $S = \frac{1}{2}(a + b + c)$.

Prism volume (Filling it up, not turning it up)

Volume is space — solid space that's measured in cubic units. Want to know the easiest figure to find the volume of? A rectangle. Well, to really be proper about this, it's actually a right rectangular prism. The area of a rectangle is length (l) times width (w). That's only two dimensions. Baby, we're working with three, so we gotta throw the height (or altitude (h), as it's formally known) into the mix. And what makes finding the volume of a right rectangular prism possible? Drum roll, please . . . $V = lwh$ (see Figure 14-4).

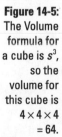
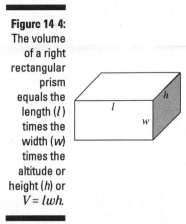

Figure 14-4:
The volume of a right rectangular prism equals the length (l) times the width (w) times the altitude or height (h) or $V = lwh$.

Here's a riddle for you: What is right, rectangular, and regular? It's a cube. The volume of a cube is the same as the volume of a rectangular prism because a square is a rectangle. For a cube, the values of l, w, and h are all the same. So you can simplify to $V = s^3$, which is read as "s cubed." (Hmmm. Wonder why.) Check out Figure 14-5.

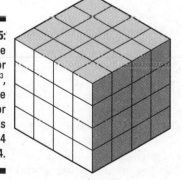

Figure 14-5:
The Volume formula for a cube is s^3, so the volume for this cube is $4 \times 4 \times 4 = 64$.

Table 14-1 lists the formulae you use to find the volume of various types of right prisms.

Table 14-1:	Volume Formulae for Right Prisms
Type of Right Prism	*Formula for Volume*
Rectangular	$V = lwh$ where l is the length of the base of the prism, w is the width of the base, and h is the altitude
Cube	$V = s^3$ where the length, width, and altitude are equal
Triangular	$V = \dfrac{lwh}{2}$ where a triangle is half a rectangle, so it has half the volume
Any right prism	$V = Bh$ where B is the area of the base and h is the altitude

A Visit to the Pyramids

Triangles again. They're everywhere in geometry. A pyramid uses triangles to form its lateral faces. But just because the lateral faces are triangular doesn't mean that the base follows suit. It can have a mind of its own.

Pyramid kinds

The base of a pyramid is a polygon. The lateral sides of a pyramid meet the base at the base edges and meet each other at the vertex of the pyramid. The base and vertex of a pyramid don't live together on the same plane. They're separated by distance. The base resides in one plane while the vertex of the pyramid resides in another plane. The farther away they are from each other, the longer the altitude (or height) is. The base determines the number of lateral faces the pyramid has. If the base is a triangle, then the result is a triangular pyramid. A square base makes a square pyramid. A *square* pyramid? Yup, that's what I said. It only means that the base is square, not the sides (see Figure 14-6).

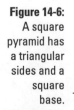

Figure 14-6:
A square pyramid has a triangular sides and a square base.

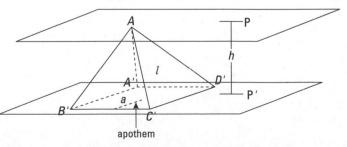

apothem

A square is a regular polygon; all the bottom lengths of the lateral sides have the same length. In addition, all the lateral edges of the pyramid in Figure 14-6 are congruent. The resulting pyramid is called a regular pyramid. The sides of the pyramid are slanted inward from the base and meet at the vertex. You obtain the height of the slanted sides of a regular pyramid via — what else? — a formula. This kind of height is called, uh, *slant height,* and only regular pyramids have it. The Slant Height formula is based on the Pythagorean Theorem ($c^2 = a^2 + b^2$ — see Theorem 6-17).

Following its predecessor, the Slant Height formula is $\ell^2 = a^2 + h^2$. I hear ya: "Wait — what's an ℓ here?" Oh, it's just used to represent the slant length of a regular pyramid. No big deal. Just a letter. You know h; it's the altitude. The altitude of a pyramid is from the vertex straight down through the center of the pyramid to the middle of the base. And a is the apothem of the base polygon (see Figure 14-7).

The *apothem* is a line segment that runs from the center of a polygon to the side of a polygon. It's a perpendicular bisector of that side (see Theorems 5-12 and 5-13).

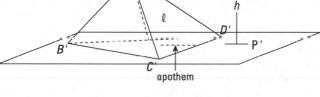

Pyramid surface area

So many sides! To get a better idea of how to calculate the surface area of a regular pyramid, peel back the lateral sides of the pyramid, like a banana, until they lie flat with the base. What do you see? Take a look at Figure 14-8. It shows five triangles and one pentagon.

Calculate ℓ, where ℓ represents the slant height of the pyramid and is obtained by using $\sqrt{a^2 + h^2}$. Get the perimeter (P) of the base polygon by adding the lengths of all the sides of the base. Now you're good to go. Plug those numbers into the formula

$$S = \frac{1}{2}\ell P$$

and out pops the lateral surface area. Hold on, though. You're not done yet. You need the area of the base polygon. It's a pentagon. The area of a regular polygon equals ½aP, where a is the apothem and P is the perimeter (see Theorem 5-14). Find that area and add it to the other one, and then you're done. You can go now.

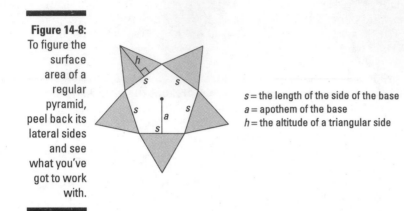

s = the length of the side of the base
a = apothem of the base
h = the altitude of a triangular side

Pyramid volume

This one's easy to keep brief. The volume of a pyramid is one-third the volume of a prism. Because the volume of a prism is $V = Bh$, where B is the area of the base polygon and h is the length of the altitude, then the volume of a pyramid is $V = \frac{1}{3}Bh$.

If you have a regular square pyramid with a base edge length of s = 4 inches and an altitude of h = 6 inches, you can find the volume by doing the following: First, get the area of the base. The base is a square, so the area is the side length squared — 4^2, or 16 square inches. So the volume is

V = ⅓ (16 square inches)(6 inches) = 32 inches cubed

Notice the *cubed* part. Squared inches multiplied by inches is indeed *cubed inches*.

Cylinders: Not Just Circles Anymore

Time to round things out a little bit. I don't want to beat around the bush, so I'll get right to the point . . . maybe. Read on.

Cylinder areas

A cylinder is like a prism, BUT (notice that this is a *big* but) the bases aren't polygons; they're circular (see Figure 14-9). You've got the same base terminology — the bases and the altitude. You've got the right circular cylinder, which stands up straight. And you've got the oblique circular cylinder, which is inclined to lean.

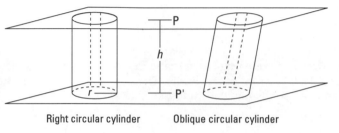

Right circular cylinder   Oblique circular cylinder

The lateral surface area of a prism is $S = hP$. But circles don't have perimeters; they have circumferences. So you need to alter the formula slightly so that it applies to a cylinder: $S = hC$, where C is the circumference. OK, that's the lateral area. Now for the total area. It's the same as for a prism: $T = S + 2B$. In other words, the total surface area equals the lateral surface area plus the area of the two bases. Remember, the bases of a right circular cylinder are circles, plain and simple. So it makes sense to change B to πr^2 because B is the area of the base and the base is a circle.

Cylinder volume

It's the same as a prism: $V = Bh$, where B is the area of the base and h is the altitude. But because you're working with a circle, I can provide you with an alternative to B. The area of a circle is πr^2 so the formula becomes $V = \pi r^2 h$.

Cones: Think Ice Cream or Party Hats

A cone is kind of a hybrid. It's like a pyramid, except it's not. It's like a cylinder, except it's not. When you think of a cone's shape, you can think of an ice cream cone or a party hat. Either way, that's the shape (see Figure 14-10).

Figure 14-10:
Cones are
the hybrid
3-D object.
They're sort
of a combo
of pyramids
and
cylinders.

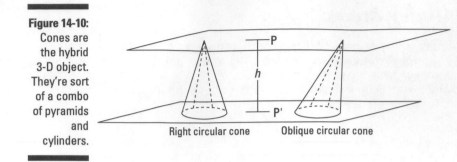

Right circular cone   Oblique circular cone

Like prisms and cylinders, the circular cone has right and oblique versions. A cone has a vertex in one plane like that of a pyramid and a base that's a circle in another plane like that of a cylinder. The cone also has a slant height like the pyramid. All solids have altitudes, so that's nothing new.

Cone areas

The lateral area is like a rounded pyramid. So, if the lateral area of a pyramid is $S = \frac{1}{2}\ell P$, then you should be able to figure out what to do to this formula to get it to work for a cone. Yep, change the P to a C. There are no perimeters with a rounded surface — just circumference. So the lateral area of a cone is $S = \frac{1}{2}\ell C$.

The total surface area of a cone is the same as a pyramid. So the area of a cone is $T = B + S$, where B is the area of the cone's only base. Because the base of a cone is a circle, $B = \pi r^2$, where r is the radius of the circle.

Cone volume

Once again, there's a parallel to the pyramid. Not literally. Just figuratively. The formula is the same: $V = \frac{1}{3}Bh$, where B for a cone is the area of a circle. This formula can be tailored slightly for a better fit for the cone: $V = \frac{1}{3}\pi r^2 h$, where r is the radius of the circle and h is the altitude.

"Sphlandering Sphema, Batman! Is That a Sphere?"

A sphere is made up of all points an equal distance from a certain point. That sounds remarkably like a circle. Oh, did I mention that these points defy confinement to one plane? A sphere is a ball that is solid, like a baseball (see Figure 14-11).

Figure 14-11: A baseball is an example of a sphere.

Sphere surface area

The surface area of sphere S with radius r is $S = 4\pi r^2$. Not much else to say about it because the origins of this formula are in calculus, and *trust me,* I'm not going there. How about an example?

If a sphere has a radius of 5 inches, then what's the surface area? $S = 4\pi(5)^2$, or 100π.

Sphere volume

The volume of a sphere is actually derived from two other volume formulae. And you may have seen 'em before. The volume of a sphere is equal to the volume of a cylinder minus the volume of two cones:

$$V = \pi r^2 h - 2\left(\frac{\pi r^3}{3}\right) \text{ where } h = 2r$$

$$= \pi r^2 (2r) - \frac{2\pi r^3}{3}$$

$$= 2\pi r^3 - \frac{2\pi r^3}{3}$$

$$= \frac{4\pi r^3}{3}$$

Part VI
The Part of Tens

The 5th Wave   By Rich Tennant

"We all know it's a pie, Helen. There's no need to pipe the equation, 3.141592653... on the top."

In this part . . .

Ah, The Part of Tens. What more can I say?

Here, you get a list of ten careers that use geometry, just in case you're thinking there can't be any real-world uses for it.

Geometry's not so bad once you get over the phobia. Just be sure to breathe and stay calm. And follow my ten tips to make your life with geometry a dream. A good dream, not the nightmare-ish kind.

Chapter 15

Ten Cool Careers That Use Geometry

I still don't know what I want to do when I grow up. Do you? Well, if you're considering any of the careers listed in this chapter, geometry is going to play a *big* part in your life. This chapter rattles off ten (OK, there are 11 — whoopee twang) careers that use geometry, but many others are available. For each career listed, I provide a brief job description and a short blurb on what aspects of geometry may be particularly handy.

Air Traffic Controller

Air traffic controllers coordinate the movement of air traffic to make sure that planes remain a safe distance from each other. The main responsibility is to control the flow of aircraft in and out of the airport using radar and the airplanes' flight plans. However, air traffic controllers do monitor all aircraft traveling in airport airspace.

Air traffic controllers use coordinates to determine the positions of airplanes under their charge. They must track an airplane's longitude, latitude, and altitude to safely manage their airspace.

Architect

Architects design buildings and other structures. They're involved in all phases of development — from designing to suit the purpose of the building to be constructed, to incorporating requirements, to budgeting. Architects draw the plans of the building and provide details about the building's appearance and its structural system. Building materials must be specified, and the building structure must comply with building codes. Architects often work with engineers, urban planners, interior designers, and landscapers to make sure that all bases are thoroughly covered.

Architects draw lots of lines and angles in their plans. Typical tools of the trade include the ruler and the protractor. Drawings are to scale, so ratios are definitely and frequently used.

Carpenter

Carpenters cut, fit, and assemble wood or other building materials in the construction of buildings, bridges, and other structures. Carpenters who work as general contractors frame walls and partitions, and they put in doors and windows. They work with blueprints and instructions. They begin by laying things out. They measure, mark, and arrange materials. They cut and shape wood, plastics, or metal by using hand or power tools. They must check the accuracy of work by using levels, plumb bobs, and framing squares. They make adjustments as necessary and assemble the pieces.

Carpenters use levels to construct parallel building supports. Corners of a building, doors, and windows must be squared.

Fashion Designer

Fashion designers design clothing and accessories. They develop a clothing line — putting together colors and materials to be worn during a season.

Fashion designers deal with materials such as fabric. They must determine the size of shapes. In laying out projects of material amounts, they use circumference, area in square yards, length, and width. Proportions are used for sizing.

Interior Designer

Interior designers plan space and furnish interiors of homes, office buildings, and commercial space. They plan additions and renovations. They develop designs and prepare working drawings with specifications for interior construction, furnishings, lighting, and lighting fixtures. They may use computer programs that allow them to lay out designs and easily make changes.

Interior designers use proportions (by scaling down an environment to create a replica) to allow them to accurately manipulate representations before any work is done. They commonly use lines and angles to determine the placement of furnishings.

Mason

Masons work with materials like bricks, tile, and stone. They plan the appearance, estimate the amount and type of materials, and figure the budget required to complete a project. Projects may include walkways, facades, chimneys, or floor surfacing.

Masons use distance and angles in their drawings that detail a project. They must establish the depth needed to make a structure safe. They use knowledge of solid geometry to estimate how much solid material is needed to fill area. They use tools like levels to check the accuracy of their work.

Mechanical Engineer

Mechanical engineers plan and design tools, engines, machines, and other mechanical equipment. They work with technical drawings that indicate the dimensions of the tools and parts to be constructed.

Mechanical engineers use rulers for distance, protractors to get the angles just right, calculators for calculations (including trigonometric functions), and compasses for those curved lines. And they use ratios to shrink a drawing so that it is drawn to scale.

Nautical Navigator

Although the innovation of GPS (global positioning satellite) devices has curtailed the use of the sextant as a navigational device, it's still important for any serious boater to know how to do *dead reckoning* — locate your position relative to a particular astronomic landmark. The angle of elevation is recorded and used in an attempt to pinpoint (or in nautical terms, triangulate) position.

Navigation in this manner uses a bit of trig and some coordinate geometry.

Surveyor

Land surveyors establish official land, airspace, and water boundaries. They write descriptions of land for deeds, leases, and other legal documents. They manage survey parties that measure distances, directions, angles between points, angles of elevation, lines, and contours of the earth's surface. Land surveyors plan fieldwork, select known survey points, and determine the precise location of important features within the survey space.

Surveyors use an instrument called a transit to determine the angles of elevation and depression. Tangents are also used, so trig is important here.

Tool and Die Maker

Tool and die makers produce tools, dies, and special holding fixtures that are used to produce a variety of products from clothing to furniture to heavy equipment. Toolmakers craft precision tools to cut, shape, and form metal and other materials. They make the jigs and fixtures needed to hold the material while they mill, turn, or grind it. Die makers construct metal forms known as dies. These forms are used to punch out or stamp materials such as aluminum. They make metal models for diecasting and for molding plastics, ceramics, and composite materials. Both tool and die makers work with blueprints and instructions. They must check the accuracy of their work and assemble parts.

Tool and die makers work a lot with trigonometric functions. They use right triangles for angular interpolation. Angles, distance, and arcs are important specifications that need to be incorporated into their product.

3-D Graphic Artist

3-D graphic artists or animators are responsible for creations that can show up at the movies. They use computer graphics programs and knowledge of solid geometry to develop a three-dimensional wire frame of the object they're working with. The artists use ratios and proportions to scale their creations, and they onlay surface textures to reflect realism and lighting effects to the surface.

Chapter 16

Ten Hot Tips to Make Geometry Easier

• •

*T*he first rule of life? Life can be difficult. Why make it more difficult than it has to be? Here are ten tips (plus a free bonus tip) that have worked for me. Hopefully, they'll work for you, too.

The second rule of life? There are no guarantees. But if you don't try, you deprive yourself of the chance to succeed. Now why would you want to do that?

Use a Clear Plastic Protractor

Tools are fun, and the dandy protractor is no exception. I like the clear plastic kind because I can see through them. That way, I can extend my angles right through the scale of the protractor. Reading angle measures is *much* easier then.

Use a Clear Plastic Ruler with Inches and Centimeters

Tools, tools, tools. Love 'em. Just as with a clear plastic protractor, with a clear plastic ruler, you can extend your lines, which makes getting their measures easier. Using a ruler with inches *and* centimeters is a good idea. Go metric, baby!

Buy Thyself a Compass

You gotta have a protractor for your angles and a ruler for your straight lines. You also need a compass — for those curved lines. I like the ones that have rulers right on them. That way, you don't have to use both the compass and a separate flat ruler when making a circle. Just pull the compass apart the distance you want and use the ruler right on it. Too cool.

Get a Good Pencil to Draw Fine Lines

You need a pencil to make accurate drawings. I recommend a technical drawing pencil with a .05 mm lead. An eraser is an important commodity, also. You won't regret getting one that has a brush. Eraser residue leaves the page easier when it's brushed away. Eraser residue always sticks to your hand anyway — use the brush.

Buy Thyself a Good Scientific Calculator

Never underestimate the power of a good scientific calculator. I'm talking about the kind with sin, cos, and tan keys. You're going to need those keys on trig days. Square root and squared keys are useful for all that triangle stuff. And fresh batteries are a good idea on test day.

Write Down Your Givens and Wants

When you're setting up to solve a problem, be it a proof or just an equation, write down everything you're given to work with even if it doesn't seem important. The smallest details can lead to the biggest revelations. After you finish with what you've been given, move on to what you want. Write that down, too.

Make Diagrams

A picture is worth a thousand words. Make a diagram with your awesome technical pencil. Try to draw things in proportion, keeping your spatial relationships intact. Mark off everything in the drawing that is in your given. If you have congruent lines, mark 'em. If you have congruent angles and parallel lines, mark 'em.

Develop a Plan of Attack

You have your given. You've written down what you want. You've drawn your diagram. Now you have to develop a plan to solve the proof. A plan of attack can be everything from which auxiliary lines you need to draw to the type of reasoning you're going to use to solve the proof. If you develop a plan before you start, it gives you direction and saves on the number of steps you'll have to take to get from the given to the prove statements.

Read Through the Statements

This suggestion works best with completed proofs — proofs actually completed by someone else, like in a book. Read through the numbered information in the Statements column. Try to figure out what the reason should be for each statement. Check to see whether you're correct. If you are, go on to the next statement. If you aren't, figure out why the reason is what it is before you proceed. Going through the steps without having to create them and just trying to understand the logic behind them is the best way to get a handle on complex proofs. Just like using training wheels is the best way to get a handle on how to ride a bike.

Apply Geometry Objects to the Real World

Apply the objects of geometry to the real world as you learn about them. Make everything a mind game. For circles, think pizza. For rectangles, think tennis courts. For spheres, think baseballs. You get the idea. Associating the information to something you already understand not only helps speed up your understanding but also improves your chances of keeping the info in your memory. There are lots of things to remember in geometry. It's time to expand your mental capacity.

Play Pool!

OK. Don't tell anyone, but this is an eleventh tip. I wanted to add this extra one so that you have a good example of how to apply geometry to the real world. Pool is all about angles. Hit the ball off one bumper at a certain angle, and it may hit another ball. Change the angle and you may hit the ball so that it rebounds from one bumper to another and sinks a solid colored ball. You may scratch or mistakenly sink the 8 ball. So play pool — angles, angles, angles!

Part VII

Appendixes

The 5th Wave   By Rich Tennant

Beyond Euclidean and Cartesian geometry, there is Ed Dubrowski geometry which proves that the volume of a sphere changes in proportion to the amount of food at an All-U-Can-Eat buffet.

In this part . . .

The material in this part is a special gift from me to you. Here, you find valuable reference material. There's a little bit of everything: tables and chairs (just kidding about the chairs); all the formulae you could ever want (or as many as I could squeeze into this part); and lists of all those wonderful postulates, theorems, and corollaries you encounter in this book.

Oh, I almost forgot — you also get a glossary, for those times when the words are on the tip of your tongue but you just can't seem to get them out.

Appendix A

Squares, Square Roots, and a Trig Table (It Jist Don't Get More Exciting Than This)

● ●

*T*his appendix provides two tables for one good reason — easy reference. Table A-1 displays the squares and square roots of the numbers 1 through 130 in case you don't have a calculator handy but need this kind of info in a jiffy. Table A-2 is a trig table (for use only with right triangles). From it, you can find an acute angle's degree measure by looking up the angle's sine, cosine, or tangent decimal value in the Sin θ, Cos θ, or Tan θ column, respectively. You can also find the decimal equivalent of an acute angle's sine, cosine, or tangent ratio, if you have the degree measure of an acute angle. The information in Table A-2 can be obtained with a scientific calculator, but if you ain't got one or your batteries have died, use the table.

Table A-1			Squares and Square Roots from 1 – 130					
x	x^2	$\sqrt{x}$	x	x^2	$\sqrt{x}$	x	x^2	$\sqrt{x}$
1	1	1.000	10	100	3.162	19	361	4.359
2	4	1.414	11	121	3.317	20	400	4.472
3	9	1.732	12	144	3.464	21	441	4.583
4	16	2.000	13	169	3.606	22	484	4.690
5	25	2.236	14	196	3.742	23	529	4.796
6	36	2.449	15	225	3.873	24	576	4.899
7	49	2.646	16	256	4.000	25	625	5.000
8	64	2.828	17	289	4.123	26	676	5.099
9	81	3.000	18	324	4.243	27	729	5.196

(continued)

Table A-1 *(continued)*

x	x^2	$\sqrt{x}$	x	x^2	$\sqrt{x}$	x	x^2	$\sqrt{x}$
28	784	5.292	56	3,136	7.483	84	7,056	9.165
29	841	5.385	57	3,249	7.550	85	7,225	9.220
30	900	5.477	58	3,364	7.616	86	7,396	9.274
31	961	5.568	59	3,481	7.681	87	7,569	9.327
32	1,024	5.657	60	3,600	7.746	88	7,744	9.381
33	1,089	5.745	61	3,721	7.810	89	7,921	9.434
34	1,156	5.831	62	3,844	7.874	90	8,100	9.487
35	1,225	5.916	63	3,969	7.937	91	8,281	9.539
36	1,296	6.000	64	4,096	8.000	92	8,464	9.592
37	1,369	6.083	65	4,225	8.062	93	8,649	9.644
38	1,444	6.164	66	4,356	8.124	94	8,836	9.695
39	1,521	6.245	67	4,489	8.185	95	9,025	9.747
40	1,600	6.325	68	4,624	8.246	96	9,216	9.798
41	1,681	6.403	69	4,761	8.307	97	9,409	9.849
42	1,764	6.481	70	4,900	8.367	98	9,604	9.899
43	1,849	6.557	71	5,041	8.426	99	9,801	9.950
44	1,936	6.633	72	5,184	8.485	100	10,000	10.000
45	2,025	6.708	73	5,329	8.544	101	10,201	10.050
46	2,116	6.782	74	5,476	8.602	102	10,404	10.100
47	2,209	6.856	75	5,625	8.660	103	10,609	10.149
48	2,304	6.928	76	5,776	8.718	104	10,816	10.198
49	2,401	7.000	77	5,929	8.775	105	11,025	10.247
50	2,500	7.071	78	6,084	8.832	106	11,236	10.296
51	2,601	7.141	79	6,241	8.888	107	11,449	10.344
52	2,704	7.211	80	6,400	8.944	108	11,664	10.392
53	2,809	7.280	81	6,561	9.000	109	11,881	10.440
54	2,916	7.348	82	6,724	9.055	110	12,100	10.488
55	3,025	7.416	83	6,889	9.110	111	12,321	10.536

x	x²	√x		x	x²	√x		x	x²	√x
112	12,544	10.583		119	14,161	10.909		126	15,876	11.225
113	12,769	10.630		120	14,400	10.954		127	16,129	11.269
114	12,996	10.677		121	14,641	11.000		128	16,384	11.314
115	13,225	10.724		122	14,884	11.045		129	16,641	11.358
116	13,456	10.770		123	15,129	11.091		130	16,900	11.402
117	13,689	10.817		124	15,376	11.136				
118	13,924	10.863		125	15,625	11.180				

Table A-2: A Trig Table (Values of Trigonometric Functions)

θ	Sin θ	Cos θ	Tan θ		θ	Sin θ	Cos θ	Tan θ
1°	.0175	.9998	.0175		18°	.3090	.9511	.3249
2°	.0349	.9994	.0349		19°	.3256	.9455	.3443
3°	.0523	.9986	.0524		20°	.3420	.9397	.3840
4°	.0698	.9976	.0699		21°	.3584	.9336	.3839
5°	.0872	.9962	.0875		22°	.3746	.9272	.4040
6°	.1045	.9945	.1051		23°	.3907	.9205	.4245
7°	.1219	.9925	.1228		24°	.4067	.9135	.4452
8°	.1392	.9903	.1405		25°	.4226	.9063	.4663
9°	.1564	.9877	.1584		26°	.4384	.8988	.4877
10°	.1736	.9848	.1763		27°	.4540	.8910	.5095
11°	.1908	.9816	.1944		28°	.4695	.8829	.5317
12°	.2079	.9781	.2126		29°	.4848	.8746	.5543
13°	.2250	.9744	.2309		30°	.5000	.8660	.5774
14°	.2419	.9703	.2493		31°	.5150	.8572	.6009
15°	.2588	.9659	.2679		32°	.5299	.8480	.6249
16°	.2756	.9613	.2867		33°	.5446	.8387	.6494
17°	.2924	.9563	.3057		34°	.5592	.8290	.6745

(continued)

Table A-2 *(continued)*

θ	Sin θ	Cos θ	Tan θ	θ	Sin θ	Cos θ	Tan θ
35°	.5736	.8192	.7002	63°	.8910	.4540	1.9626
36°	.5878	.8090	.7265	64°	.8988	.4384	2.0503
37°	.6018	.7986	.7536	65°	.9063	.4226	2.1445
38°	.6157	.7880	.7813	66°	.9135	.4067	2.2460
39°	.6293	.7771	.8098	67°	.9205	.3907	2.3559
40°	.6428	.7660	.8391	68°	.9272	.3746	2.4751
41°	.6561	.7547	.8693	69°	.9336	.3584	2.6051
42°	.6691	.7431	.9004	70°	.9397	.3420	2.7475
43°	.6820	.7314	.9325	71°	.9455	.3256	2.9042
44°	.6947	.7193	.9657	72°	.9511	.3090	3.0777
45°	.7071	.7071	1.0000	73°	.9563	.2924	3.2709
46°	.7193	.6947	1.0355	74°	.9613	.2756	3.4874
47°	.7314	.6820	1.0724	75°	.9659	.2588	3.7321
48°	.7431	.6691	1.1106	76°	.9703	.2419	4.0108
49°	.7547	.6561	1.1504	77°	.9744	.2250	4.3315
50°	.7660	.6428	1.1918	78°	.9781	.2079	4.7046
51°	.7771	.6293	1.2349	79°	.9816	.1908	5.1446
52°	.7880	.6157	1.2799	80°	.9848	.1736	5.6713
53°	.7986	.6018	1.3270	81°	.9877	.1564	6.3138
54°	.8090	.5878	1.3764	82°	.9903	.1392	7.1154
55°	.8192	.5736	1.4281	83°	.9925	.1219	8.1443
56°	.8290	.5592	1.4826	84°	.9945	.1045	9.5144
57°	.8387	.5446	1.5399	85°	.9962	.0872	11.4301
58°	.8480	.5299	1.6003	86°	.9976	.0698	14.3007
59°	.8572	.5150	1.6643	87°	.9986	.0523	19.0811
60°	.8660	.5000	1.7321	88°	.9994	.0349	28.6363
61°	.8746	.4848	1.8040	89°	.9998	.0175	57.2900
62°	.8829	.4695	1.8807	90°	1.0000	.0000	

Appendix B

Important Formulae: A Quickie Guide

• •

Granted, the formulae presented in this book are unforgettable. However, just in case your memory gets a little fuzzy, you don't have to go thumbing through the pages of this book to find all the formulae in it. This appendix provides you one-stop service for the most important ones.

The General Measurements of Angles

Table B-1 summarizes the general measurements of some important angle types.

Table B-1	The General Measurements of Angles
Angle	*Formula*
Complement (c) of $\alpha°$	$c = 90° - \alpha°$
Supplement (s) of $\alpha°$	$s = 180° - \alpha°$
Sum measure (S) of the interior angles of a triangle	$S = 180°$
Sum measure (S) of the interior angles of a quadrilateral	$S = 360°$
Sum (S) of the exterior angles of a polygon	$S = 360°$
Sum (S) of the interior angles of an n-gon	$S = 180°(n-2)$

The Measurements of Angles in Circles

Table B-2 provides a summary of the measurements of the angles that are found in and about a circle.

Table B-2	The Measurements of Angles in Circles
The Location of the Angle's Vertex	*The Measure of the Angle Equals . . .*
At the center of the circle (central angle)	. . . the measure of the intercepted arc.
On the circle (inscribed angle)	. . . one-half the measure of the intercepted arc.
Inside the circle (two intersecting chords)	. . . one-half the sum of the two measures of the intercepted arcs.
Outside the circle (two intercepting tangents, or secants, of the tangent-secant)	. . . one-half the difference of the two measures of the intercepted arcs.
Inside a semicircle (angle inscribed in a semicircle)	. . . 90°.
Opposite angles inside an inscribed quadrilateral	. . . 180° minus the measure of one of the angles (equals the measure of the opposite angle).

Trigonometry Formulae

Here are the formulae for sine, cosine, and tangent.

The sine ratio is $\sin \theta = \dfrac{opposite}{hypotenuse}$

The cosine ratio is $\cos \theta = \dfrac{adjacent}{hypotenuse}$

The tangent ratio is $\tan \theta = \dfrac{opposite}{adjacent}$

Area Formulae

Table B-3 gives you a handy reference to locate the area formula for a variety of different polygons, as well as some parts of geometric figures.

Table B-3	Area Formulae
Area	*Formula*
Area of a triangle	$A = \frac{1}{2}bh$ where b measures the base and h, the altitude
Area of a triangle using perimeter	$A = \sqrt{s(s-a)(s-b)(s-c)}$, where $s = \frac{1}{2}(a+b+c)$ and a, b, c are lengths of the sides
Area of a right triangle	$A = \frac{1}{2}(ab)$ where a and b are measures of the two legs
Area of an equilateral triangle	$A = (s^2/4)\sqrt{3}$ where s is the measure of a side
Area of a parallelogram	$A = bh$ where b measures the base and h, the altitude
Area of a rectangle	$A = bh$ or $A = lw$ where l measures the length and w, the width
Area of a square	$A = s^2$ where s is the measure of a side
Area of a trapezoid	$A = \frac{1}{2}h(b_1 + b_2)$ where b_1 and b_2 are measures of the bases and h, the altitude
Area of a rhombus or a kite	$A = \frac{1}{2}(d_1)(d_2)$ where d_1 and d_2 are measures of the diagonals
Area of a regular polygon	$A = \frac{1}{2}aP$, where a is the apothem and P is the perimeter
Area of a circle	$A = \pi r^2$
Area of a sector of a circle	$A = \frac{n}{360}\pi r^2$, where r is the radius and n is the measure of the central angle or intercepted arc
Area of a minor segment of a circle	$A = \left(\frac{n}{360}\pi r^2\right) - \frac{1}{2}bh$, where r is the radius, n is the measure of is the measure of the intercepted arc, b is the base of the triangle, and h is its altitude.

Circle and Line-Segment Relationships

Figure B-1 shows the relationship of the lengths of two chords. Figure B-2 shows the relationship between the lengths of a tangent and a secant. And Figure B-3 shows the relationship of the lengths of two secants.

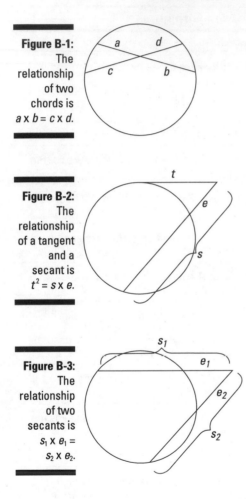

Figure B-1:
The relationship of two chords is $a \times b = c \times d$.

Figure B-2:
The relationship of a tangent and a secant is $t^2 = s \times e$.

Figure B-3:
The relationship of two secants is $s_1 \times e_1 = s_2 \times e_2$.

Perimeter, Circumference, and Length of Arc Formulae

Table B-4 shows the arc perimeter, circumference, and length formulae.

Table B-4	Perimeter, Circumference, and Length of Arc Formulae
Quantity	*Formula*
Perimeter (P) of a triangle	$P = a + b + c$ where a, b, c are lengths of the sides of the triangle
Perimeter (P) of a rectangle	$P = 2b + 2h$ where b measures the width and h is the height
Perimeter (P) of a square	$P = 4s$ where s is the measure of a side
Perimeter (P) of a regular polygon with n sides	$P = (n)(s)$ where n is the number of sides of the polygon and s is the measure of one of those sides
Circumference (C) of a circle	$C = 2\pi r$ or $C = \pi d$ where r is the measure of the radius and d is the diameter
Length (ℓ) of an arc	$\ell = \dfrac{n}{360} C$ where n is the degree measure of the arc or the measure of the central angle and C is the circumference of the circle.

Right Triangle Formulae

Table B-5 contains the formulae for the oh-so-special right triangle.

Table B-5	Right Triangle Formulae	
Model	**Quantity**	**Formula**
	Pythagorean Theorem	$c^2 = a^2 + b^2$
	Leg opposite 30° angle	$b = \frac{1}{2}c$
	Leg opposite 60° angle	$b = \frac{1}{2}c\sqrt{3}$ or $a = b\sqrt{3}$
	Leg opposite 45° angle	$b = \frac{1}{2}c\sqrt{2}$ or $b = a$
	Altitude of equilateral triangle	$h = \frac{1}{2}s\sqrt{3}$
	Side of equilateral triangle	$s = \frac{2}{3}h\sqrt{3}$
	Side of square	$s = \frac{1}{2}d\sqrt{2}$
	Diagonal of square	$d = s\sqrt{2}$

Coordinate Geometry Formulae

Table B-6 contains the analytic geometry formulae for use with the coordinate grid.

Table B-6	Coordinate Geometry Formulae
Quantity	*Formula*
Distance	$d = \sqrt{(x_2 - x_1)^2 + (y_2 - y_1)^2}$ where x and y are coordinates of points
Midpoint	$M = \left(\dfrac{x_1 + x_2}{2}, \dfrac{y_1 + y_2}{2} \right)$ where x and y are coordinates of points
Slope	$m = \dfrac{y_2 - y_1}{x_2 - x_1}$ where x and y are coordinates of points
Equation of a line	Slope-intercept form: $y = mx + b$ where b is the y-intercept, m the slope, and x and y are coordinates of points
	Point-slope form: $y_2 - y_1 = m(x_2 - x_1)$ where m the slope, and x and y are coordinates of points
Equation of a circle	$(x - h)^2 + (y - k)^2 = r^2$ where h and k are coordinates of the center point of the circle, x and y represent a point on the circle, and r is the radius.

Surface Area Formulae

Table B-7 shows the total area or the surface area formulae.

Table B-7	Total Surface Area Formulae
Quantity	**Formula**
Right prism	$T = S + 2B$, where $S = hP$ or the lateral area and B is the base area
Regular pyramid	$T = S + B$, where $S = \frac{1}{2}\ell P$ or the lateral area and B is the base area
Right circular cylinder	$T = 2\pi rh + 2\pi r^2$ where r is the radius of the base and h is the height or altitude
Right circular cone	$T = \pi r\ell + \pi r^2$ where r is the radius of the base and ℓ is the slant length
Sphere	$T = 4\pi r^2$ where r is the length of the radius

Volume Formulae

Table B-8 shows the volume formulae for some solid figures.

Table B-8	Volume Formulae
Quantity	**Formula**
Right prism	$V = Bh$ where B is the area of the base polygon and h is the length of the altitude
Cube	$V = s^3$ where s is the length of any edge
Pyramid	$V = \frac{1}{3}Bh$ where B is the area of the base polygon and h is the length of the altitude
Right circular cylinder	$V = \pi r^2 h$ where r is the radius of the base and h is the altitude
Right circular cone	$V = \frac{1}{3}\pi r^2 h$ where r is the radius of the base and h is the altitude
Sphere	$V = \frac{4}{3}\pi r^3$ where r is the radius

Appendix C

Postulates, Theorems, and Other Delectables

• •

Yep, here it is — all in one big, huge list. This appendix contains all the postulates, theorems, and corollaries covered in this book. I left the numbering on them so you can trace them back to the chapter content from which they came. I threw in some principles and rules from Chapter 3, too. I think they're important and shouldn't be overlooked. Besides, in other geometry books, you may see these principles listed as postulates.

The Postulates

Postulate 1-1: Two points determine a line.

Postulate 2-1: The whole quantity is equal to the sum of all of its parts.

Postulate 2-2: If ray $\overrightarrow{YN}$ is interior to angle *XYZ*, then the measure of angle *XYN* and the measure of angle *NYZ* equals the measure of angle *XYZ*. (Postulate 2-1 is kind of a general statement. When it is specifically applied to angles, as it is here in Postulate 2-2, it can be referred to as the Angle Addition Postulate.)

Postulate 2-3: Given a line and a point not on that line, exactly one parallel line may be drawn through the given point.

Postulate 2-4: Two lines in the same plane either run parallel to each other or intersect.

Postulate 2-5: If two lines are crossed by a transversal, and the alternate interior angles are congruent, then the lines are parallel.

Postulate 2-6: If two parallel lines are crossed by a transversal, then their alternate interior angles are congruent.

Postulate 2-7: If two lines are perpendicular to a transversal, then these two lines are parallel to each other.

Postulate 5-1: If a polygon encloses smaller, non-overlapping regions within its perimeter, then the area of that polygon is equal to the sum of the areas of the enclosed regions.

Postulate 6-1: Two triangles are congruent if two sides and the included angle of one triangle are congruent to the corresponding parts of the other triangle ($SAS \cong SAS$).

Postulate 6-2: Two triangles are congruent if two angles and the included side of one triangle are congruent to the corresponding parts of the other triangle ($ASA \cong ASA$).

Postulate 6-3: Two triangles are congruent if the three sides of one triangle are congruent to the corresponding sides of the other triangle ($SSS \cong SSS$).

Postulate 6-4: Two triangles are congruent if two consecutive angles and the nonincluded side of one triangle are congruent to the corresponding parts of the other triangle ($AAS \cong AAS$ or $SAA \cong SAA$).

Postulate 6-5: Two right triangles are congruent if the lengths of the two legs of one triangle are congruent to the legs of the other triangle (leg-leg).

Postulate 6-6: Two right triangles are congruent if the corresponding leg and hypotenuse of one triangle are congruent to those of the other triangle (hypotenuse-leg).

Postulate 6-7: Two right triangles are congruent if an acute angle of one triangle is congruent to the corresponding acute angle of the other triangle and the hypotenuses are the same length (hypotenuse-angle).

Postulate 6-8: Two right triangles are congruent if an acute angle and its adjacent leg of one triangle are congruent to the corresponding parts of the other triangle (adjacent leg-angle).

Postulate 6-9: Two right triangles are congruent if an acute angle and its opposite leg of one triangle are congruent to the corresponding parts of the other triangle (opposite leg-angle).

Postulate 8-1: Two circles are considered congruent if and only if their radii are congruent.

Postulate 8-2: In the same circle or congruent circles, chords of equal lengths cut off equal arcs.

Postulate 8-3: Except for the point of tangency, all points on a tangent lie outside the circle.

Postulate 8-4: In the same circle or congruent circles, arcs of the same degree have the same length.

Postulate 9-1: Any two quantities can have one of three relationships: The first quantity is equal to the second. The first quantity is greater than the second. The first quantity is less than the second.

Postulate 9-2: Given three quantities, if the first is greater than the second and the second is greater than the third, then the first is greater than the third.

Postulate 9-3: The whole quantity is greater than any one of its parts.

Postulate 9-4: A quantity can be substituted for another of equal value in an inequality.

Postulate 9-5: If the same quantity is added to unequal quantities, the quantities are still unequal in the same order.

Postulate 9-6: If unequal quantities are added to unequal quantities of the same order, then the sums are unequal in the same order. You may see this postulate referred to as the Addition Property of Inequality.

Postulate 9-7: If equal quantities are subtracted from unequal quantities, the quantities are still unequal in the same order.

Postulate 9-8: If unequal quantities are subtracted from equal quantities, then the differences are unequal in the opposite order.

Postulate 9-9: If unequal quantities are multiplied by the same positive number, the products are unequal in the same order.

Postulate 9-10: If unequal quantities are multiplied by the same negative number, the results are unequal in the opposite order.

Postulate 9-11: If unequal quantities are divided by the same positive number, the quotients are unequal in the same order.

Postulate 9-12: If unequal quantities are divided by equal negative quantities, the quotients are unequal in the opposite order.

Postulate 9-13: Equal positive integral powers and equal positive integral roots of unequal positive quantities are unequal in the same order.

The Theorems

Theorem 1-1: If two lines intersect, then they do so at exactly one point.

Theorem 2-1: Given a line segment $\overline{AD}$ with points B and C between endpoints A and D, if $\overline{AB} = \overline{CD}$, then $\overline{AC} = \overline{BD}$.

Theorem 2-2: Two angles are congruent if they are both right angles.

Theorem 2-3: Two angles are congruent if they are both straight angles.

Theorem 2-4: Two angles are congruent if they are complements of the same angle or a congruent angle.

Theorem 2-5: Two angles are congruent if their complements are congruent.

Theorem 2-6: Two angles are congruent if they are supplements of the same angle or a congruent angle.

Theorem 2-7: Two angles are congruent if their supplements are congruent.

Theorem 2-8: Two angles are congruent if they are vertical angles.

Theorem 2-9: A perpendicular line segment is the shortest segment that can be drawn from a point to a line.

Theorem 2-10: If a pair of corresponding angles is congruent, then the lines are parallel.

Theorem 2-11: If a pair of alternate exterior angles is congruent, then the lines are parallel.

Theorem 2-12: If a pair of same-side interior angles is supplementary, then the lines are parallel.

Theorem 2-13: If lines are parallel, then any pair of corresponding angles is congruent.

Theorem 2-14: If lines are parallel, then any pair of alternate exterior angles is congruent.

Theorem 2-15: If lines are parallel, then any pair of same-side interior angles is supplementary.

Theorem 2-16: If two parallel lines are cut by a transversal, any pair of angles is either congruent or supplementary.

Theorem 5-1: The total number of unique diagonals (D) for a polygon of n sides is $D = \frac{1}{2} n(n - 3)$.

Theorem 5-2: The sum of the measures of the interior angles of a triangle is 180°.

Theorem 5-3: The sum of the measures of the interior angles of a quadrilateral is 360°.

Theorem 5-4: If a polygon has n sides, the formula for the sum of its interior angles is $S = 180°(n - 2)$.

Theorem 5-5: In a regular polygon, the measure of an interior angle is equal to $180°(n - 2)/n$.

Theorem 5-6: The sum of the exterior angles of a regular polygon is 360°.

Theorem 5-7: In a regular polygon, the measure of each exterior angle is $360°/n$.

Theorem 5-8: Radii of a regular polygon bisect the interior angle.

Theorem 5-9: Central angles of a regular polygon are congruent.

Theorem 5-10: Central angles of regular polygons with equal sides are congruent.

Theorem 5-11: The measure of a central angle in a regular polygon is equal to 360 divided by the number of sides of the polygon.

Theorem 5-12: An apothem of a regular polygon bisects the central angle (determined by the side) to which it's drawn.

Theorem 5-13: An apothem of a regular polygon is a perpendicular bisector to the side it's drawn to.

Theorem 5-14: The formula for the area of a regular polygon is $A = \frac{1}{2}ap$, where a is the apothem and p is the perimeter.

Theorem 6-1: The three medians of a triangle are concurrent.

Theorem 6-2: The point of concurrency of the medians of a triangle is, from any vertex, two-thirds the distance from that vertex to the midpoint of the opposite side.

Theorem 6-3: The midline of a triangle is parallel to the third side.

Theorem 6-4: The midline is half as long as the third side of the triangle.

Theorem 6-5: The altitudes of a triangle (or the lines containing the altitudes) are concurrent.

Theorem 6-6: In a given triangle, the product of the length of any side and the length of the altitude drawn to that side is equal to the product of the length of any other side and the altitude drawn to that side.

Theorem 6-7: The perpendicular bisectors of the sides of a triangle are concurrent at a point equidistant from any vertex of the triangle.

Theorem 6-8: The angle bisectors of a triangle are concurrent at a point equidistant from every side of the triangle.

Theorem 6-9: If two sides of a triangle are congruent, then the angles opposite those sides are congruent.

Theorem 6-10: If two angles of a triangle are congruent, then the opposite sides are congruent.

Theorem 6-11: The bisector of the vertex angle of an isosceles triangle divides the triangle into two congruent triangles.

Theorem 6-12: The altitude of an equilateral triangle is the square root of 3 divided by 2 multiplied by the length of a side of the triangle.

Theorem 6-13: If the square of the length of the longest side is less than the sum of the squares of the lengths of the other two shorter sides, then the triangle is acute.

Theorem 6-14: If the square of the length of the longest side is greater than the sum of the squares of the lengths of the other two shorter sides, then the triangle is obtuse.

Theorem 6-15: A triangle may have at most one obtuse angle.

Theorem 6-16: The length of the median to the hypotenuse of a right isosceles triangle is equal to one-half the length of the hypotenuse.

Theorem 6-17: For a right triangle, the sum of the squares of the lengths of the legs equals the length of the square of the hypotenuse. That is, $a^2 + b^2 = c^2$, which is otherwise known as the Pythagorean Theorem.

Theorem 6-18: In a triangle, if the square of the length of the longest side equals the sum of the squares of the lengths of the other two shorter sides, then the triangle is a right triangle. That is, if $c^2 = a^2 + b^2$, then you can conclude that the triangle is a right triangle. This is the Converse of the Pythagorean Theorem.

Theorem 6-19: The area of a triangle is $A = \frac{1}{2}bh$.

Theorem 7-1: If both pairs of opposite sides of a quadrilateral are congruent, the quadrilateral is a parallelogram.

Theorem 7-2: If both pairs of opposite angles of a quadrilateral are congruent, the quadrilateral is a parallelogram.

Theorem 7-3: A diagonal of a parallelogram divides the parallelogram into two congruent triangles.

Theorem 7-4: Consecutive angles of a parallelogram are supplementary.

Theorem 7-5: The area of parallelogram is equal to the product of the length of the base and the length of its corresponding altitude.

Theorem 7-6: All angles in a rectangle are right angles.

Theorem 7-7: The diagonals of a rectangle are congruent.

Theorem 7-8: If a quadrilateral is equilateral, it is a rhombus.

Theorem 7-9: The diagonals of a rhombus are perpendicular.

Theorem 7-10: A parallelogram is a rhombus if the diagonals bisect the vertex angles.

Theorem 7-11: The two diagonals of a rhombus form four congruent triangles.

Theorem 7-12: The area of a rhombus is one-half the product of the lengths of the two diagonals, or $A = \frac{1}{2}((d_1)(d_2))$.

Theorem 7-13: A square is an equilateral quadrilateral.

Theorem 7-14: A square is a rhombus with one right angle.

Theorem 7-15: The measure of a diagonal in a square is the length of any side multiplied by the square root of 2.

Theorem 7-16: The base angles of an isosceles trapezoid are congruent.

Theorem 7-17: The diagonals of an isosceles trapezoid are congruent.

Theorem 7-18: The area of a trapezoid is equal to one-half the product of the length of the altitude multiplied by the sum of the bases.

Theorem 8-1: All radii of a given circle or congruent circles are congruent.

Theorem 8-2: In the same circle or congruent circles, chords that are equidistant from the center of the circle are equal. The converse is also true: Equal chords are equidistant from the center of the circle.

Theorem 8-3: The diameter is twice the distance of the radius.

Theorem 8-4: All diameters of a given circle are congruent.

Theorem 8-5: A radius is perpendicular to a tangent at the point of tangency. That is, at the point of tangency, the measure of the angle that's formed by the intersection of a radius and a tangent is 90°.

Theorem 8-6: If two circles are tangential to each other, then the line of centers connecting the two circles is perpendicular to their common tangent.

Theorem 8-7: If two chords intersect in a circle, the product of the length of the segments of one chord equals the product of the length of the segments of the other chord.

Theorem 8-8: An angle formed by the intersection of a tangent and a chord is one-half the measure of the intercepted arc.

Theorem 8-9: If two secants form an angle external to a circle, then the product of the length of one secant and the length of its internal segment is equal to the product of the length of the other secant and the length of its internal segment.

Theorem 8-10: If two tangents form an angle external to a circle, then the tangents are equal.

Theorem 8-11: The circumference of a circle is 360°.

Theorem 8-12: The length of an arc of a circle is equal to $2\pi r \times \left(\dfrac{n}{360\, degrees} \right)$.

Theorem 8-13: The measure of an inscribed angle is one-half the measure of its intercepted arc.

Theorem 8-14: The measure of an angle exterior to a circle that is formed by two secants is one-half the difference of its intercepted arcs.

Theorem 8-15: If a diameter is perpendicular to a chord, then it bisects the chord and its arcs. The bisector creates congruent segments and congruent arcs.

Theorem 8-16: In a circle, parallel chords cut off equal arcs.

Theorem 8-17: The area of a circle = πr^2.

Theorem 8-18: The sides of a circumscribed polygon are tangential to the inscribed circle.

Theorem 8-19: The length of the radius of a circle inscribed in an equilateral triangle is one-third of the length of the altitude of the triangle.

Theorem 8-20: All sides of an inscribed polygon are chords of the circle.

Theorem 8-21: If a circle is circumscribed about a polygon, then the interior polygon is inscribed in the circle. So it can be referred to as an inscribed polygon. If the inscribed polygon is a quadrilateral, the opposite angles of the inscribed quadrilateral are supplementary.

Theorem 8-22: If a parallelogram is inscribed within a circle, then it is a rectangle.

Theorem 10-1: In a proportion, the cross-products are equal. That is, the product of the means and the product of the extremes are equal.

Theorem 10-2: If the products of two pairs of numbers are equivalent, then either pair can be made the means and the other pair made the extremes in a proportion.

Theorem 10-3: The ratios on both sides of a proportion can be inverted and still remain in proportion.

Theorem 10-4: Either the means or the extremes of a proportion can change position in the equation, and the ratios still remain in proportion.

Theorem 10-5: In a proportion, adding the second proportional to the first proportional and adding the fourth proportional to the third proportional results in an equivalent proportion.

Theorem 10-6: In a proportion, subtracting the second proportional from the first proportional and subtracting the fourth proportional from the third proportional results in an equivalent proportion.

Theorem 10-7: A line that is parallel to one side of a triangle divides the other two sides of the triangle proportionally.

Theorem 10-8: If a line divides two sides of a triangle proportionally, then it is parallel to the third side.

Theorem 10-9: Two triangles are similar if two pairs of corresponding angles are congruent.

Theorem 10-10: Two triangles are similar if one pair of corresponding congruent angles has proportional lengths as the sides of the angles.

Theorem 10-11: Two triangles are similar if their corresponding sides are proportional.

Theorem 11-1: The distance between two points that have the same ordinate is the absolute value of the difference of their abscissas.

Theorem 11-2: The distance between two points that have the same abscissa is the absolute value of the difference of their ordinates.

Theorem 11-3: The distance between two points — (x_1, y_1) and (x_2, y_2) — is determined by the formula

$$D = \sqrt{(x_2 - x_1)^2 + (y_2 - y_1)^2}$$

Theorem 11-4: The slopes of two parallel lines are equal.

Theorem 11-5: Parallel lines have equal slopes.

Theorem 11-6: If two lines are perpendicular, then the product of their slopes is −1.

Theorem 11-7: If two lines are perpendicular, then the products of the slopes of the two lines is equal to −1. The slopes of those lines are negative reciprocals of each other.

Theorem 11-8: A circle is defined by the equation

$$r^2 = (x - h)^2 + (y - k)^2$$

Theorem 12-1: A locus of points equidistant from two given points is a perpendicular bisector of the line connecting those two points.

Theorem 12-2: A locus of points equidistant from two given intersecting lines is the bisector of any angles formed by these lines.

Theorem 12-3: The locus of a set of points that are an equal distance — on both sides — from a given line is a pair of lines; each of them is equidistant from and parallel to the given line, and the lines are also parallel to each other. The given line is midway between the two lines.

Theorem 12-4: The locus of points that are a given distance from a given point is a circle whose center is the given point, and the length of the radius is the given distance.

Theorem 12-5: The locus of points that are a given (equal) distance from the outside of a circle is a circle outside the given circle and concentric with it.

Theorem 12-6: The locus of points whose abscissa is a constant value is a vertical line parallel to the y-axis.

Theorem 12-7: The locus of points whose ordinate is a constant value is a horizontal line parallel to the x-axis.

The Corollaries

Corollary 6-1: An angle bisector of the vertex angle of an isosceles triangle is a perpendicular bisector of the base of the triangle.

Corollary 6-2: The bisector of the vertex angle of an isosceles triangle bisects the base.

Corollary 6-3: An equilateral triangle is also equiangular.

Corollary 6-4: The three angles of an equilateral triangle each have a measure of 60° (see Theorem 5-2).

Corollary 6-5: A triangle may have at most one right angle (see Theorem 5-2).

Corollary 6-6: The acute angles of a right triangle are complementary (see Theorem 5-2).

The Principles and Rules

Reflexive principle: $a = a$

Symmetric principle: If $a = b$, then $b = a$

Transitive principle: If $a = b$ and $b = c$, then $a = c$

Substitution principle: If $x = 20$ and $x + y = 30$, then $20 + y = 30$

Addition rule: If $a = b$ and $c = d$, then $a + c = b + d$

Subtraction rule: If $a = b$ and $c = d$, then $a - c = b - d$

Multiplication rule: If $a = b$, then $2a = 2b$

Division rule: If $a = b$ and $c = d$, then $a / c = b / d$

Roots rule: If $a = b$, then $\sqrt{a} = \sqrt{b}$

Powers rule: If $a = 7$, then $(a)^2 = (7)^2$

Glossary

*T*his glossary provides you with all (well, most) of the words you may need to come to terms with in your exposure to geometry. I even alphabetized them for you all by myself. Brain power, *yeah!*

acute angle: An angle with a measure of more than 0° but less than 90°.

acute triangle: A triangle in which all three angles have a measure of less than 90°.

addition principle: A rule stating that if equal quantities are added to equal quantities, their sums are also equal.

adjacent: Next to or neighboring.

adjacent angles: Two angles in the same plane that share a common side and vertex but have no common interior points.

alternate interior angles: Interior angles on either side of a transversal.

altitude: A line that's representative of the height of a figure.

angle: The union of two rays or two line segments with a common endpoint.

angle bisector: A line that splits an angle into two equal angles. Also, a ray that has an endpoint at the vertex of an angle and divides the angle into two equal, or congruent, angles.

apothem: A line that runs from the center of a polygon straight into a flat side of the polygon.

arc: a curved part or segment of a circle that is defined by two points and all the points in between.

arrowhead: A quad with one angle greater than 180° and two pairs of adjacent congruent sides.

axiom: A statement or basic assumption that's accepted as being true without proof. It's basically the same as a postulate (depending on whom you talk to).

central angle: An angle with a vertex at the center point of a circle that has radii as sides.

centroid: The point where all three medians of a triangle meet.

chord: A line segment that divides a circle into two segments: a minor segment and a major segment.

chord-chord angle: An angle formed by the intersection of two chords in a circle.

circle: A set of points in a plane that are the same (fixed) distance from a specific point.

circumscribed circle: A circle that is about a polygon in such a way that each vertex of the polygon touches the circle at exactly one point.

closed figure: A shape that distinctly separates an internal region from an external one.

collinear points: Points that lie on the same line.

commensurable: Units that can be converted to the same unit of measure.

common tangent: A line that is tangent to two circles.

complementary angles: A pair of angles whose measures add up to 90°.

concurrent lines: Three or more lines that exist at the same time.

cone: A solid geometric shape with a circular base in one plane tapering to a point in another plane.

congruent: Having equal measures.

consecutive: Occurring one right after another.

coordinate geometry: A branch of geometry that uses ordered pairs mapped on a grid.

coordinate plane: A grid that contains a means for determining both horizontal and vertical indicators of a location.

corresponding angles: Angles on the same side of a transversal and in the same position relative to the lines that the transversal crosses.

cosine ratio: In a right triangle, the ratio of the length of the side adjacent to an acute angle over the hypotenuse.

cotangent ratio: In a right triangle, the ratio of the length of the adjacent side to a given acute angle over the length of the opposite side.

CPCTC: A rule stating that corresponding parts of congruent triangles are congruent.

cylinder: A solid geometric figure whose circular bases are in separate planes.

deductive reasoning: A method of reasoning in which conclusions about the information are drawn only after a step-by-step process involving the logical progression of statements leads to a conclusion.

degree: The most commonly used unit of measurement for an angle.

determined: Having just enough information so that one line satisfies the given.

diameter: A chord that passes through the center point of a circle and runs from one side of the circle to the other. (The formula for calculating the diameter is $D=2r$, where r is the radius.)

division principle: A rule stating that if equal quantities are divided by an equal quantity (other than zero), their quotients are equal.

equiangular triangle: A triangle in which the measure of each of the three angles is equal, or 60°.

equilateral triangle: A triangle with three equal sides.

exterior angle: An angle that lies outside a figure.

external tangent: A common tangent that doesn't cross the line of centers.

externally tangent circles: Two circles whose centers lie on the opposite side of a common tangent.

geometria: A Latin word from which the term *geometry* originated. *Ge* means "earth" and *metrein* means "measure."

incommensurable quantities: Quantities that can't be compared because there's no common unit between them to form a ratio (in other words, quantities that have different units, like miles per gallon).

indirect reasoning: A method of reasoning in which you assume that what you want to prove is false.

inductive reasoning: A method of reasoning in which you make an organized attempt to test a theory based on experimentation and observation. The conclusion has a high degree of probability for occurrence.

inscribed angles: Angles formed by two chords with the vertex as a point on the circle.

inscribed circle: A circle within a polygon where each of the sides of the polygon is tangent to the circle.

interior angle: An angle that is inside a figure and whose sides are also sides of the polygon.

internal tangent: A common tangent that crosses through the line of centers.

internally tangent circles: Two circles whose centers lie on the same side of a common tangent.

intuition: A method of reasoning based on that "I just know it" feeling. It's not really based on anything else. You jump to a conclusion without a thorough analysis of the facts.

isosceles triangle: A triangle with two equal sides.

kite: A quad with two pairs of adjacent congruent sides and with no angle measuring more than 180°.

line: A set of points that extends infinitely in both directions.

line of centers: A line that's drawn from the center of one circle to the center of another.

line segment: A part of a line that has two endpoints that mark its finite length.

locus: The set of all points and only those points that satisfies a given condition or set of conditions. The plural form is *loci*.

median: A line drawn from a vertex to the midpoint of the opposite side.

midline: A line drawn from the midpoint of one side of a triangle to the midpoint of a second side and that's parallel to the base.

midpoint: a point on a line segment or an arc that separates it into two congruent parts.

minute: A unit of angle measurement. Sixty minutes equal one degree.

multiplication principle: A rule stating that if equal quantities are multiplied by an equal quantity, their products are equal.

oblique geometric solid: A geometric solid in which the altitude of the shape is not perpendicular to its bases.

obtuse angle: An angle with a measure of more than 90° and less than 180°.

obtuse triangle: A triangle in which the measure of one of the angles is greater than 90°.

ordered pair: A location on a coordinate plane given as (x, y).

origin: The point located at the intersection of the x- and y-axes with a location indicated by the ordered pair of $(0, 0)$.

overdetermined: Having too much information in that just one line could never satisfy the given.

parallel lines: Lines that don't intersect or have any points in common.

parallelogram: A quad with opposite parallel sides.

perpendicular bisector: A line that intersects another line, forming two 90° angles and splitting the segment into two equal parts.

plane: An infinite flat surface that has no depth or boundaries.

point: An indicator of a definite location or position. A point, though, has no width, no length, and no depth.

point of intersection: The location at which two lines cross.

point of tangency: The one location at which a tangent contacts a circle.

polygon: A closed shape bounded by at least three sides, each side being a straight line. The word *polygon* has a Greek origin, meaning "many angled."

polyhedra (or polyhedrons): Solid geometric shapes bounded by multiple planes. The terms have a Greek origin, meaning "multisided."

postulate: A statement or basic assumption that's accepted as being true without proof.

prism: A solid whose congruent bases lie on separate planes.

protractor: A tool that can be used to measure or make an angle.

pyramid: A geometric shape that has triangles as its lateral faces, or sides.

quadrilateral: A polygon with exactly four sides.

radius: A line that extends from the center to the side of a figure. The plural form is *radii*.

rate: The quotient of two incommensurable quantities.

ratio: The quotient of two numbers where the denominator is not zero.

ratio of similitude: A ratio of the measure of any two corresponding sides of similar triangles (or polygons).

ray: A part of a line that has only one endpoint and extends infinitely in one direction.

rectangle: A parallelogram with a right angle. Its nickname among geometry geeks is *rect*.

reflexive principle: A rule stating that a quantity is equal (or congruent) to itself.

rhombus: A parallelogram with equal sides.

right angle: An angle with a measure of 90°.

right geometric solid: A geometric solid in which the lateral edges or altitude of the shape are perpendicular to its base edges.

right triangle: A triangle that has an angle with a measure of 90°.

scalene triangle: A triangle with no equal sides or angles.

secant: A line (or segment or ray) that intersects a circle at two distinct points.

second: A unit of angle measurement. Sixty seconds equal one minute.

sector of the circle: The region of the circle bounded by the radii that create the arc.

similar polygons: Polygons in which all pairs of corresponding angles are congruent and their three pairs of corresponding sides are in proportion.

sine ratio: In a right triangle, the ratio of the length of the side opposite a given acute angle over the hypotenuse.

slope: The amount of vertical change of a line relative to the amount of its horizontal change.

sphere: A solid made up of all points that are an equal distance from a certain point.

square: An equilateral quadrilateral.

square roots principle: A rule stating that positive square roots of equal quantities are equal.

squares principle: A rule stating that the squares of equal quantities are equal.

straight angle: An angle that has a measure of 180°.

substitution principle: A rule stating that a quantity may be substituted for its equivalent in an expression.

subtraction principle: A rule stating that equal quantities may be subtracted from both sides of an equation.

supplementary angles: A pair of angles whose measures add up to 180°.

symmetric principle: A rule stating that an equal quantity can be reversed.

tangent: A line (or segment or ray) that intersects (or touches) a circle at exactly one point.

tangent ratio: In a right triangle, the ratio of the lengths of the opposite side over the adjacent side to a given acute angle.

tangent-chord angle: An angle formed by a chord and a tangent where the vertex of the angle is at the point of tangency.

theorem: A statement that has been proven to be true.

transitive principle: A rule stating that if two quantities are equal to the same quantity, then the two quantities are equal to each other.

transversal: A line that crosses two or more lines at different points.

trapezoid: A quad with one pair of parallel sides and mutually perpendicular diagonals.

triangle: A polygon with exactly three sides.

trigonometry: An area of geometry that uses the properties of right triangles to determine measurements. *Trig* for short.

undefined terms: Words that can't be defined in simpler terms, so they're described instead of defined.

underdetermined: Having too little information in that more than one line satisfies the given.

vertex: In a polygon, the point at which two lines intersect. The plural form is *vertices.*

vertex angle: In an isosceles triangle, the angle formed by the two congruent legs of an isosceles triangle; the angle opposite the base.

vertical angles: Angles that are formed by intersecting lines and are opposite each other.

volume: The amount of cubic units contained within a three-dimensional geometric shape.

***x*-axis:** The horizontal axis in a coordinate plane that has values that increase from left to right.

***x*-coordinate:** The number that represents the location of a point relative to the *x*-axis.

***y*-axis:** The vertical axis in a coordinate plane that has values that increase from bottom to top.

***y*-coordinate:** The number that represents the location of a point relative to the *y*-axis.

Index

• *Q* •

FOR DUMMIES®

A world of resources to help you grow

HOME, GARDEN & HOBBIES

Feng Shui FOR DUMMIES
A Reference for the Rest of Us!
0-7645-5295-3

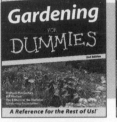

Gardening FOR DUMMIES
A Reference for the Rest of Us!
0-7645-5130-2

Guitar FOR DUMMIES
A Reference for the Rest of Us!
0-7645-5106-X

Also available:

Auto Repair For Dummies
(0-7645-5089-6)

Chess For Dummies
(0-7645-5003-9)

Home Maintenance For Dummies
(0-7645-5215-5)

Organizing For Dummies
(0-7645-5300-3)

Piano For Dummies
(0-7645-5105-1)

Poker For Dummies
(0-7645-5232-5)

Quilting For Dummies
(0-7645-5118-3)

Rock Guitar For Dummies
(0-7645-5356-9)

Roses For Dummies
(0-7645-5202-3)

Sewing For Dummies
(0-7645-5137-X)

FOOD & WINE

Cooking FOR DUMMIES
A Reference for the Rest of Us!
0-7645-5250-3

Cookies FOR DUMMIES
A Reference for the Rest of Us!
0-7645-5390-9

Wine FOR DUMMIES
A Reference for the Rest of Us!
0-7645-5114-0

Also available:

Bartending For Dummies
(0-7645-5051-9)

Chinese Cooking For Dummies
(0-7645-5247-3)

Christmas Cooking For Dummies
(0-7645-5407-7)

Diabetes Cookbook For Dummies
(0-7645-5230-9)

Grilling For Dummies
(0-7645-5076-4)

Low-Fat Cooking For Dummies
(0-7645-5035-7)

Slow Cookers For Dummies
(0-7645-5240-6)

TRAVEL

Italy FOR DUMMIES
A Travel Guide for the Rest of Us!
0-7645-5453-0

Hawaii FOR DUMMIES
A Travel Guide for the Rest of Us!
0-7645-5438-7

Las Vegas FOR DUMMIES
A Travel Guide for the Rest of Us!
0-7645-5448-4

Also available:

America's National Parks For Dummies
(0-7645-6204-5)

Caribbean For Dummies
(0-7645-5445-X)

Cruise Vacations For Dummies 2003
(0-7645-5459-X)

Europe For Dummies
(0-7645-5456-5)

Ireland For Dummies
(0-7645-6199-5)

France For Dummies
(0-7645-6292-4)

London For Dummies
(0-7645-5416-6)

Mexico's Beach Resorts For Dummies
(0-7645-6262-2)

Paris For Dummies
(0-7645-5494-8)

RV Vacations For Dummies
(0-7645-5443-3)

Walt Disney World & Orlando For Dummies
(0-7645-5444-1)

Available wherever books are sold. Go to www.dummies.com or call 1-877-762-2974 to order direct.

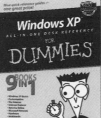

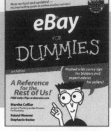

FOR DUMMIES®

The advice and explanations you need to succeed

SELF-HELP, SPIRITUALITY & RELIGION

Sex FOR DUMMIES
0-7645-5302-X

Parenting FOR DUMMIES
0-7645-5418-2

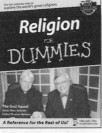

Religion FOR DUMMIES
0-7645-5264-3

Also available:

The Bible For Dummies
(0-7645-5296-1)

Buddhism For Dummies
(0-7645-5359-3)

Christian Prayer For Dummies
(0-7645-5500-6)

Dating For Dummies
(0-7645-5072-1)

Judaism For Dummies
(0-7645-5299-6)

Potty Training For Dummies
(0-7645-5417-4)

Pregnancy For Dummies
(0-7645-5074-8)

Rekindling Romance For Dummies
(0-7645-5303-8)

Spirituality For Dummies
(0-7645-5298-8)

Weddings For Dummies
(0-7645-5055-1)

PETS

Puppies FOR DUMMIES
0-7645-5255-4

Dog Training FOR DUMMIES
0-7645-5286-4

Cats FOR DUMMIES
0-7645-5275-9

Also available:

Labrador Retrievers For Dummies
(0-7645-5281-3)

Aquariums For Dummies
(0-7645-5156-6)

Birds For Dummies
(0-7645-5139-6)

Dogs For Dummies
(0-7645-5274-0)

Ferrets For Dummies
(0-7645-5259-7)

German Shepherds For Dummies
(0-7645-5280-5)

Golden Retrievers For Dummies
(0-7645-5267-8)

Horses For Dummies
(0-7645-5138-8)

Jack Russell Terriers For Dummies
(0-7645-5268-6)

Puppies Raising & Training Diary For Dummies
(0-7645-0876-8)

EDUCATION & TEST PREPARATION

Spanish FOR DUMMIES
0-7645-5194-9

Algebra FOR DUMMIES
0-7645-5325-9

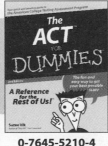

The ACT FOR DUMMIES
0-7645-5210-4

Also available:

Chemistry For Dummies
(0-7645-5430-1)

English Grammar For Dummies
(0-7645-5322-4)

French For Dummies
(0-7645-5193-0)

The GMAT For Dummies
(0-7645-5251-1)

Inglés Para Dummies
(0-7645-5427-1)

Italian For Dummies
(0-7645-5196-5)

Research Papers For Dummies
(0-7645-5426-3)

The SAT I For Dummies
(0-7645-5472-7)

U.S. History For Dummies
(0-7645-5249-X)

World History For Dummies
(0-7645-5242-2)